TIMBER and the FOREST SERVICE

DEVELOPMENT OF WESTERN RESOURCES

The Development of Western Resources is an interdisciplinary series focusing on the use and misuse of resources in the American West. Written for a broad readership of humanists, social scientists, and resource specialists, the books in this series emphasize both historical and contemporary perspectives as they explore the interplay between resource exploitation and economic, social, and political experiences.

John G. Clark, University of Kansas, General Editor

TIMBER and the FOREST SERVICE

David A. Clary

Published by the University Press of Kansas (Lawrence, Kansas 66045), which was organized by the Kansas Board of Regents and is operated and funded by Emporia State University, Fort Hays State University, Kansas State University, Pittsburg State University, the University of Kansas, and Wichita State University

Library of Congress Cataloging-in-Publication Data
Clary, David A.
Timber and the Forest Service.
(Development of western resources)
Bibliography: p.
Includes index.
1. United States. Forest Service—History.
2. Forest management—United States—History. 3. Timber
—United States—History. I. Title. II. Series.
SD143.C538 1986 333.75'15'0973 86-15762
ISBN 0-7006-0314-X

Printed in the United States of America
10 9 8 7 6 5 4 3 2 1

To the memory of Oscar O. Winther

CONTENTS

LIST OF ILLUSTRATIONS

PHOTOGRAPHS

MAPS

PREFACE

In 1905 the Bureau of Forestry, a small element in the Department of Agriculture, became the Forest Service. In addition to its previous assignment of advising the nation on forestry, the new organization assumed charge of the National Forest System, then in its infancy. Over the following half-century or more, the energetic agency enjoyed public approval, even applause, of a kind seldom awarded federal bureaucracies. Beginning in the 1950s, however, and increasingly in the next two decades, the Forest Service suffered savage condemnation, as its staunchest supporters became its bitterest critics. How did such things come about?

They happened because the Forest Service was and is different from other bureaucracies, which operate on a body of policy that continually adjusts to conditions in the world. The Forest Service, however, has something more like a religion. From the beginning it has perceived itself as fulfilling a sacred mission to provide wood to the world in order to avert the evils of a "timber famine."

That steadfast devotion to an ideal explains much that is commendable about the agency and its people—their esprit de corps, their sense of mission, their dedication, their record of honesty. It also explains how the Forest Service got itself into so much trouble during the 1960s and 1970s. How else to account for the organization's unbending adherence, in the face of contrary facts and changing conditions, to principles and practices adopted decades ago? How else to understand the Forest Service's inability to listen to the views of others? When the American people demanded other values than timber from the national forests, the Forest Service answered with preaching.

Over the years, the Forest Service developed a number of devices to keep others from bending its dedication. The most important was "multiple use," under the banner of which the Forest Service had everyone's interests at heart, serving as arbiter of competing claims. Repeatedly the Forest Service has been able to establish the terms of an argument, then lose it. That happened with multiple use, which the public discovered was something the Forest Service preached but did not practice. The Service, however, always emerged from controversies feeling bruised but correct, believing firmly that its first mission is the production of timber.

This is, accordingly, a study of a unique bureaucratic culture. It is the story of how a dedicated group of people endeavored to serve the public interest as they defined it and of what happened when the public defined its interest differently. This book is, therefore, not a condemnation of the Forest Service, as that would be unwarranted. It is, however, an attempt to examine a case of public service wherein the servant believed firmly that it knew better than the public what the public really wanted. There are no villains or heroes in this history, merely people trying to do the right thing.

Although writing is supposed to be a solitary task, no author of history can work without assistance. I acknowledge especially the scores of people in the Forest Service and in the National Archives and Records Service who assisted in plumbing tons of documents and publications, as did Alice Wickizer and her staff of the Indiana University Library. I also must express gratitude to the Forest Service itself, for which, in 1983, I prepared a technical and administrative history of its timber-management programs. That project produced a considerable amount of useful information for the present study. In addition, it suggested the need for something beyond its original scope—an interpretive history of the Forest Service's bureaucratic and psychological relation to timber and how this relationship affected the Service's relations with society. Space does not permit listing them all, but Forest Service people who deserve special notice for their advice during that study include Dennis M. Roth, Frank Harmon, George Weyermann, Richard Millar, Ira J. Mason, Homer Hixon, Don Morris, Glen Jorgensen, Walter H. Lund, and Alfred Wiener.

I am also grateful for the advice of several outstanding authorities on forestry and forest history, including Marion Clawson, Resources for the Future; Thomas R. Cox, Department of History, San Diego State University; the late Harley H. Thomas, consulting forester; John G. Clark, Department of History, University of Kansas; William D. Rowley, Department of History, University of Nevada; and Richard W. Behan, School of Forestry, Northern Arizona University. In addition, especially helpful were Harold K. Steen, Ronald J. Fahl, Mary Elizabeth Johnson, and Richard C. Davis of the Forest History Society. And last, I say thanks to "Cousin Brent"; he knows why.

All of the foregoing helped to make this work possible, and each deserves credit for what is of value in it. Any shortcomings, however, are entirely my own. In the final analysis, I thank above all my wife,

Beatriz, without whose patience and encouragement I could not do anything.

David A. Clary
Greene County, Indiana
December 1985

Alice looked round her in great surprise. ''Why, I do believe we've been under this tree all the time! Everything's just as it was!''

''Of course it is,'' said the Queen: ''what would you have it?''

''Well, in our *country,'' said Alice, still panting a little, ''you'd generally get to somewhere else—if you ran very fast for a long time, as we've been doing.''*

''A slow sort of country!'' said the Queen. ''Now, here, *you see, it takes all the running* you *can do, to keep in the same place. If you want to get somewhere else, you must run at least twice as fast as that!''*

—Lewis Carroll

PROLOGUE

It behooves every forester to find justification for his own art and for his own existence in the answer to the fact that natural supplies are waning and are not being replaced as fast as consumed.

—Bernhard E. Fernow (1902)

Telling the Iroquois Indians to beware of the English colonists in the 1750s, the Marquis Duquesne warned, "The forest falls before them as they advance, and the soil is laid bare so that you can scarce find the wherewithal to erect a shelter for the night."[1] He voiced what would become a recurrent nightmare of American history: if the nation consumed its forests without thought of the future, it should one day find itself without the timber upon which its civilization depended.[2]

The American forest had seemed limitless at first, its disappearance inconceivable. Besides, the forest was a barrier, harboring forbidding dangers and covering land that ought to be farmed. The forest represented everything civilization was not; perhaps a modern nation would be better off without it. During the nineteenth century, however, a different attitude emerged. The rising price of firewood caused concern at first, but that was abated by the general adoption of stoves in place of fireplaces.[3] Presently it appeared that wood for other uses could become scarce unless future supplies were secured. As farms and towns spread over the continent, even ordinary construction materials might some day be hard to find.

Civilization itself might be threatened if that should come to pass. In 1864, George Perkins Marsh wrote *Man and Nature,* the "fountainhead" of the conservation movement.[4] Marsh attributed the fall of lost civilizations to the wasteful use of their natural resources, especially forests. He believed it was not necessary for a nation to collapse from such a cause. Foresight would ensure prosperity: "The sooner a natural wood is brought into the state of an artificially regulated one, the better it is for all the multiple interests which depend on the wise administration of this branch of public economy."[5] The message was clear: the nation that destroys its forests destroys itself; take care of natural resources, "civilize" the forest, or perish.

Several influential Americans took Marsh's message to heart, as the timber frontier crossed the continent, leaving in its wake denuded landscapes and human misery—Marsh's prescription for the fall of nations. The United States might face a "timber famine," a time when forest resources would be gone forever. Public documents, one after the other, estimated the nation's timber supplies at from 856 to 2,500 billion board feet and predicted that they would be exhausted within twenty to seventy years.[6] A growing body of opinion held that the federal government should do something about the situation.

In 1891, Congress authorized the president to set aside forest reserves on the public domain.[7] That did not ensure the "wise use" of the public forests or even protect them, however. Accordingly, in 1897, Congress appropriated funds to regulate the federal forests and offered broad directions for management.[8] "For the purpose of preserving the living and growing timber and promoting the younger growth on national forests," the secretary of the Interior could sell and supervise the harvest of certain types of timber. Congress also provided that local residents could have the free use of timber and stone on the national forests. A reserve could be established "to improve and protect the forest within the boundaries, or for the purpose of securing a continuous supply of timber for the use and necessities of the citizens of the United States." The legislators believed that they had averted the horror of a "timber famine," so that Americans in the future would have "the wherewithal to erect a shelter."

CHAPTER ONE

The National Forests and the Struggle for Conservation

Forestry, as men responsible for property must view it, is conceived to be not a fixed and complete science, but a finely balanced art, with a utilitarian aim.

—Austin Cary (1914)

The forest reserves seemed a modest palliative for something as terrible as a "timber famine." They totaled only a little over 30 million acres in 1897, confined mostly to high ground in the Pacific Coast states and in the central and northern Rocky Mountains. Their origins were often dubiously linked to the name "forest reserves," because many of them had been established to protect watersheds. The most prevalent economic use was grazing, not timbering.

There were people, however, who had visions of a greater future for the reserves. They were in two overlapping categories—foresters and Progressives—and they were determined to establish a system of protected forests commensurate with what they believed the nation required. The reserves increased to 63 million acres by early 1905, and before July of that year President Theodore Roosevelt expanded the total area to 85.7 million acres. He and his successor, William Howard Taft, left behind about 187 million acres of protected federal forest land when they turned the government over to Woodrow Wilson in 1913. Purchase of eastern forest lands had begun at modest levels by that time, but Wilson contracted the system slightly. The area was about 182 million acres when he left office in 1921.[1]

The increase and adjustment of the National Forest System paralleled a struggle to define its purpose and place in the national forest-products economy. Proponents of federal conservation must develop a mechanism for forest management—a bureaucracy—and then infuse that organization with a body of working knowledge and a sense of purpose. Principles must be formulated and then tested in the real world of politics. People and ideas confronted one another in a struggle to define a national role for the forests—and for the Forest Service. The result, as always in a democratic society, was a series of compromises. That meant, in the view of the most ardent conservationists, that the

The National Forest System in 1898—the forest reserves created by Presidents Benjamin Harrison and Grover Cleveland. In this year Gifford Pinchot became chief of the Division of Forestry.

task was only partly completed by the time Wilson left office. To the more practical minded, at least a working system was in place. The Forest Service that they created, however, was a very peculiar institution, with a mind all its own.

FORESTERS, PROGRESSIVES, AND THE FORESTS

Congress authorized the forest reserves because people believed that a "timber famine" was a real and dangerous prospect. There was also growing concern over timber theft on the public domain. The later expansion of the forest system, however, reflected the zeal of groups and individuals who feared the timber famine and thought they knew how to prevent it.

The authority to manage the forest reserves included only general guidance, but it was most specific on the subject of timber. The reserves should be carefully used for material benefit, but in ways that would permit the continued extraction of timber at rates not to exceed the growth of new timber. That objective, soon to be known as "sustained yield," followed European precedents. Just how European principles

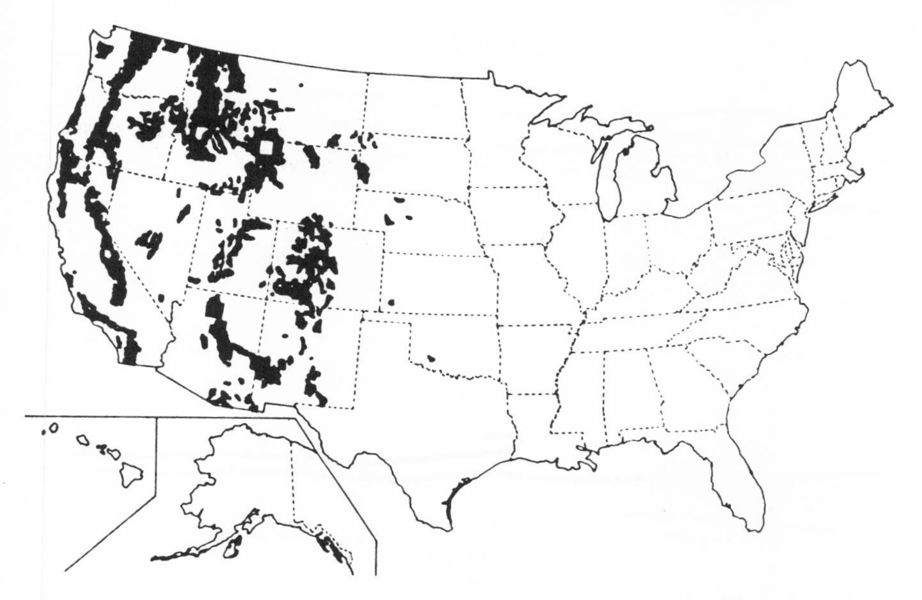

The National Forest System in 1907. President Theodore Roosevelt greatly expanded the national forest system under the Forest Reserve Act of 1897.

were to be transferred into American terms remained to be seen, however. That duty would fall to people called foresters.[2]

Administrative and political details required attention first. The federal forestry program was divided between the departments of the Interior and of Agriculture. The forest reserves fell under the jurisdiction of the General Land Office in the Interior Department. That agency had many other jobs to perform, and it had a shortage of personnel technically trained in forestry. In the Agriculture department, however, a staff of forestry experts had been growing since 1876. From one man, the organization grew successively to a "Division" and then to the Bureau of Forestry; by the turn of the century, it was the best-known body of forest authorities in the country. Already in the business of advising the public generally on forest management, the bureau eagerly took on the role of technical adviser in the General Land Office's administration of the forest reserves.

Given the belief of the bureau's leaders—Bernhard Eduard Fernow and his successor, Gifford Pinchot—that "demonstration" forestry (the technically correct management of selected woodlands as good examples for the rest of America) was of great importance, inevitably the Bureau of Forestry soon coveted the forest reserves as proving grounds for its own theories. The foresters got their wish in 1905, with the concurrence of the Department of the Interior. Administration of the reserves was transferred to the Department of Agriculture, in the particular charge of their own organization—now renamed the Forest Service. In 1907 the forest reserves were renamed "national forests," Pinchot's way of suggesting that the resources where to be used, not locked away.[3]

From the start, the progress of the national forests was joined with that of forestry. They were rather like fraternal twins, born of a peculiar mixture of Romanticism and Utilitarianism in the nineteenth century, their characters forged by the turmoils of the Progressive Era in the first years of the twentieth. It seemed that neither could exist without the other, as the national forests absorbed the energies of the majority of the young profession. It would be some time before it became apparent that the purpose of all forestry and all the purposes of national forests were not always identical. For the moment, the purpose of both was to ensure a continuing flow of timber. There had not been much forestry in the United States before there were national forests, however. As one of the profession's pioneers observed, the word *forestry* was absent from American dictionaries as late as 1880. The few technically competent foresters had no forests to manage, and as a result, "forestry in the United States was chiefly propaganda." The foremost of the first generation pointed out that in America there were "no past masters of the art."[4]

The pioneer American foresters looked for instruction to Europe, where forestry had a long history. Their principal guidebook was the five-volume compendium by Sir William Schlich, *A Manual of Forestry*. "Some of it," said an early Forest Service man, "was a straight translation out of the German and the rest of it was based on German forestry texts." It was, in short, the highest authority, and it was treated with reverence. "There is no forester in America," ventured the *Journal of Forestry* in 1927, "who is not familiar with Schlich's name even if he is not the owner of that splendid encyclopedic manual of forestry which bears his name." That same year, Forest Service Chief William B. Greeley opined, "There is no irreverence in the line of the old Yale Forest School song, 'Our notes are thick with old Bill Schlich.' "[5]

European forestry, developed in countries that had long been occupied and in which populations and industry were growing but native resource bases were not, husbanded timber resources for future use. It therefore averted the threat of a "timber famine" by applying strict technical and political principles. The early literature of American foresters reflected a strong European influence. In one of the first American texts, Henry S. Graves predicted that "in the long run, the application of forestry in this country will resemble very closely that in

Gifford Pinchot as governor of Pennsylvania. Founder of the Forest Service, he often carried his championship of forest conservation to the extreme (courtesy Forest History Society).

Europe, with such modifications as are required by the peculiar characteristics of our species and climate.'' In another book, he wrote, ''I have drawn freely from the experience of European foresters.''[6]

Graves was not alone. ''This book,'' admitted A. B. Recknagel in a text published in 1913, ''does not pretend to present any original theories of Forest Organization, but merely the best of European efforts along this line adapted to the present needs of American foresters.'' For a while there seemed to be little reason to look elsewhere than Europe, whatever the forestry topic. ''The literature of Forest Finance,'' H. H. Chapman observed in 1914, ''is largely of German origin, and has been developed in the latter half of the nineteenth century by numerous German and French authorities.'' James W. Toumey drew ''freely'' in 1916 on ''foreign literature, particularly that of Germany and France . . . for the principles underlying the practice'' of nurseries.[7]

American foresters were fascinated by the accumulated wisdom of European, especially German, forestry. They soon perceived, however, that the practices of the Old World could not simply be transferred to the New. American foresters must develop a body of American principles, applicable to American circumstances. As they did that, they also consciously separated themselves from things foreign, including even those Germans who helped to found American forestry. As Pinchot put it later: "We distrusted them and their German lack of faith in American forestry"; "As an American, I thought I might succeed where they had failed." That was no mere statement of hindsight. As early as 1899, Pinchot averred: "It would be easily possible to secure Germans or other foreigners, but a considerable experience has convinced me that, except in rare cases . . . the attempt to use foreign-born men trained abroad is not likely to succeed." When, in the following year, the Pinchot family endowed a forestry school at Yale, their announced purpose was to produce "American foresters trained by Americans in American ways for the work ahead in American forests."[8]

The United States was not Europe. Species of trees, soils, and climates on this side of the Atlantic were sufficiently different to require an independent body of knowledge. Equally important, the American political system and free enterprise were not susceptible to the dictates possible in the Old World; public acceptance must be earned, not demanded. Moreover, although European countries had long been settled, America was still filling in its landscape with a human population, and the country had vast forest resources yet untouched. The question was whether American forests should follow the history of European forests—that is, be subjected to "devastation" and then careful management—or should receive management before "devastation" occurred.

Immigrant foresters from Germany, however, first realized that the New World required new approaches. Carl Alwin Schenck observed in 1899: "We cannot import German forestry unless we import German conditions. . . . In this country, at least 85 percent of all woodland is owned by private individuals who cannot possibly be compelled to manage their forests for the general welfare when such management interferes with the owners' financial views."[9]

Bernhard Fernow agreed. European forests had been managed intensively for generations, he noted. Many American forests would require improvement work for years to come. Pinchot's aide Henry Graves echoed: "A sustained yield, an allotment of the forest into divisions, a permanent road system, the accessibility of all parts of the forest at one time, fire lines, improvement cuttings, and the like, which are usually considered a necessary part of forest management, each must, in many instances be given up as impracticable for the present."[10]

New people were more important than old ideas in American forestry. The new must be willing to learn from the old, however. Among the old European authorities, Sir Dietrich Brandis—among other distinctions, the father of forest conservation in the British Empire in India—exerted no less an influence than that of Schlich. Brandis was the intellectual father of many early American foresters, but if he had spoken to no one other than his student Gifford Pinchot, the effect would have been significant. "Brandis," claimed Pinchot's biographer, "had a more profound influence on Gifford both as a forester and as an individual than anyone except his family and Theodore Roosevelt." Pinchot's stamp is still apparent in the Forest Service, and there is a distinct flavor of Sir Dietrich's influence as well.[11]

Certain pioneers of American forestry stand out. First among them was Bernhard Eduard Fernow. A native of Germany who had been trained there in forestry, he arrived in the United States in 1876 to attend a meeting of the American Forestry Association. He remained to marry an American woman, became a citizen, and embarked upon a distinguished career in his adopted land. After working in industry, he served as head of the Department of Agriculture's Division of Forestry from 1886 to 1898. During that period he firmly implanted in American forestry the idea that a supply of wood was fundamental to civilization; his annual reports included involved calculations of the precise cubic footage of wood essential to a modern nation. He next became a pioneer in the academic world, serving as dean of new forestry schools at Cornell University and the University of Toronto. As editor of the *Forestry Quarterly* and its successor, the *Journal of Forestry,* from 1903 to 1923, Fernow exerted a remarkable influence on the development of his profession, publishing some 250 articles and bulletins and two books during his lifetime. He is regarded by many today as the father of American forestry.[12]

The most formidable figure in early American forestry was Gifford Pinchot. Born to wealth and raised in an atmosphere of social responsibility and activism, Pinchot first became interested in forestry while he was a student at Yale University in the 1880s. In 1889 he began a thirteen-month whirlwind education in Europe under Brandis's tutelage and at a forestry college in France. He returned to the United States in 1891, hung out his shingle as a consulting forester, and promptly received a commission from the Phelps-Dodge Company. The next year he became the forester for George W. Vanderbilt's Biltmore estate in North Carolina. Although Pinchot spent only three years part-time at Biltmore, the project was the nation's first "demonstration" forestland, and it gave Pinchot some professional credentials and a claim to being America's first trained forester.

Pinchot was more at home on the political field than in the woods, however. Forest conservation as social reform became his crusade. After serving on the National Forest Commission and as a special forestry agent for the Department of the Interior, in 1898 he succeeded Fernow at the Division of Forestry. He imbued his staff with a remarkable esprit de corps. A confidante of President Theodore Roosevelt's, Pinchot was for many years a powerhouse in political Progressivism. By his energy and persuasiveness he achieved the establishment of the Forest Service and its accession of the national forests in 1905. His zeal was also his undoing: President William Howard Taft fired him for insubordination in 1910. Pinchot went on to political success in Pennsylvania, serving as commissioner of forestry and twice as governor of that state. The crusade for conservation was never far from his mind, however. He remained a fighter to the end (he died in 1946), championing his cause, defending his Forest Service from its critics, always attacking those whom Roosevelt branded ''malefactors of great wealth.''[13]

Pinchot changed the conservation movement, turning it into a moral crusade that divided people too easily into camps of nobles and villains. That earned him many enemies as well as devoted admirers, but it helped to ensure that almost every difference of opinion on conservation matters would become emotional and heartfelt. One of his lifelong friends observed that characteristic of Pinchot's ''advocacy of conservation, is [his] emphasis on its moral as well as its material aspects. To [Pinchot] the moral issues involved in the equitable distribution of natural resources as between present and future generations, and in the prevention of monopoly in their ownership and utilization, are matters of prime importance.'' That played upon a contradiction in Pinchot's personality. ''Always direct, forthright and decisive, he is ordinarily the soul of kindness and consideration; but a puritanical conscience can, if he thinks the occasion demands, make him ruthless in sacrificing his friends as well as himself in behalf of a just cause.'' In what he believed was a ceaseless battle between the people and industry, Pinchot gave no quarter.[14]

Austin Cary, who also held a claim to being America's ''first'' forester, saw things somewhat differently. A native of Maine, he studied biology and entomology at Johns Hopkins and Princeton and received his A.B. and M.A. degrees from Bowdoin College, where he taught in the late 1880s. He worked thereafter as a surveyor and land estimator in New England, where he produced the first of many scores of technical publications. Becoming interested in forestry, Cary visited Europe several times, falling under the spell of Dietrich Brandis and supplementing his studies with inspections of American forests. His employment by Berlin Mills Company in New Hampshire gave Cary a claim to the honor of being the first industrial forester in the United States. Cary taught forestry at Yale and Harvard during the first decade of the twentieth century. He was a member of Maine's delegation to the

Governors' Conference on Conservation in 1908 and later became superintendent of forests in New York.

Cary's distrust of ideologues and his firm faith in the ability of private industry to adopt conservation practices caused Pinchot to deny Cary an appointment in the Forest Service. After Pinchot's departure in 1910, however, Cary assumed the position of "logging engineer," whence he began more than two decades of salesmanship for forest conservation among private owners and forest industries. The general adoption of forestry practices in New England and the South has been credited largely to Cary's influence; his personal approach and public opposition to federal regulation combined to earn him an audience among industrialists.[15]

Another student of Brandis's who worked in the United States was Carl Alwin Schenck. A native of Darmstadt, Germany, Schenck obtained a Ph.D. degree in forestry from the University of Giessen. He worked for the forestry service of Hesse-Darmstadt before he migrated to the United States to succeed Pinchot as the forester at Biltmore, a job he got on Brandis's recommendation. With Vanderbilt's backing, Schenck founded the Biltmore Forest School in 1898. From that pulpit and by using the estate as a demonstration field, he promoted practical, workaday forestry. Schenck had support from industry (accordingly, he fell out with Pinchot fairly early), but he considered it insufficient in terms of both money and the number of lumbermen's sons who were sent to the school. His textbooks enjoyed a wide circulation, largely because of their practicality, as opposed to the ideology, theory, and "policy" emphasized in much of the forestry literature of the time.

Schenck's time in America was cut short, however. That was a result of events that had nothing to do with forestry but that in the end exerted a great influence on its development in the United States: the First World War erupted in 1914, and Schenck returned home to fulfill his reserve commission and earn a wound in battle. Few Americans understood how Schenck could live in this country for nearly two decades while retaining his German citizenship and then fighting in a war that they blamed on German "militarism." Schenck remained alienated from most American foresters for several years; some perhaps never did forgive him for his German patriotism.[16]

The events that drew Schenck home also spelled the death of any remaining desire to base American forestry on the German model. Everything German was discredited, and foresters, no less than other Americans, were persuaded of the superiority of New World ways over those of the quarrelsome Old.

Americans, as it happened, had quarrels of their own, which had begun well before Schenck departed. The early history of the national

Forest destruction: the aftermath of wasteful industrial logging in the Northwest during the 1920s. The fire danger in such conditions was extreme (courtesy Forest History Society).

forests was mixed thoroughly with a larger national debate on forest conservation. Before the role of the national forests could be defined, an encompassing policy on conservation had to be propounded and sold to the public, in particular the timber industry. The latter was not an easy task, as the attitude of many lumbermen made clear. "Originally," an industry spokesman said in 1905, "forestry advocates were of two classes, either sentimentalists or technicists; the latter being trained in the forest methods of the old European countries where conditions were entirely different from those that obtained in the United States. [They] proposed the impossible."[17]

The attitudes that "forestry advocates" held toward the timber industry were equally clear. For decades before the passage of the 1897 legislation they had painted a picture of a migrating juggernaut, consuming forests and leaving wastage and misery in its wake. Unless the abuses of the big industries were halted, they averred, America would face a "timber famine" in the future. In actuality, they largely addressed phenomena that had almost passed away. Chief among them was the "lumberman's frontier," as historian Frederick Jackson Turner described it in the 1890s. Turner interpreted the settlement of the

Forest destruction: the effects of fire's consuming slash and debris left on the ground after wasteful logging. Such fires could consume top soil and potential new growth, leaving a moonscape where the growth of a new forest was dubious. This area, photographed by Samuel T. Dana in 1915, later became part of the Cochitopa National Forest in Colorado (courtesy Forest History Society).

American continent as a series of waves, characterized by different economic activities—there were the fur traders' frontier, the farmers' frontier, and so on. Each represented a step in the evolution from wilderness to civilization. The so-called lumbermen's frontier played its part midway on the evolutionary scale, arriving after traders and soldiers had driven out the Indians, clearing the land for agriculture to follow, and providing raw material for the more extensive development behind agriculture.[18]

Turner's succession of frontiers was too neat to accord with reality, however. There had been a lumbermen's frontier, but not necessarily before or after other frontiers. Nor did it or any other frontier move in orderly fashion down the decades, as Turner suggested. Large-scale lumbering occurred from colonial times on, when the economy demanded it, and accelerated when both demand and technology made acceleration possible. Sometimes the loggers cleared land that farmers plowed behind them. Elsewhere, firewood cutters and even farmers preceded the industrial loggers. The logging of virgin stands of timber first became a grand phenomenon after the Civil War, and for a period it reflected the worst excesses that "forestry advocates" attributed to the entire industry. Large financial combines formed in the timber industry, as in many other industries. Their capital permitted the opening of timbered lands that had previously been inaccessible. Logging railroads moved into wooded areas, paralleled by large crews of sawyers and increasingly productive milling machinery. The product of the forests was shipped over improved transportation to growing cities and towns, as America was still building mostly with wood.

By the turn of the century the lumberman's frontier had worked its way through the Great Lakes states and the South and had entered areas that had previously been by-passed, such as the Appalachian Mountains, or those that had previously been untapped, such as the Pacific Northwest. Especially in the Great Lakes region, the nature of the industry in the latter part of the nineteenth century made logging destructive, all-consuming, and heedless of the future. The industry was mostly migratory, and it left behind denuded areas, plagued by fires, soil erosion, and unemployment.

That had begun to change by the time the new century dawned. The economics of the industry still required heavy logging to support immediate returns, with little thought to future forest growth. And certainly, much of the industry was still characterized by large operations. But the trusts and combines that had formerly dominated the industry had fallen apart, and small operators accounted for a large part of the production. Moreover, the opportunity for private ownership of virgin timber was on the decline. Much of the remaining untouched forest land was in public ownership, and the government had made it clear that it would not tolerate "cut out and get out" logging on its property. On the other hand, the amount of commercial forest still available to private operators was enough to tide the industry over for many years; a collision between federal foresters and private exploiters of public timber was not necessarily imminent.

Finally, the economy of the timber industry had also changed. Domestic production of timber products stood at about 7.2 billion cubic feet in 1900. It peaked at 9.2 billion cubic feet a decade later, and then began to decline. By 1920, production averaged less than 8 billion cubic feet a year, at which figure it stayed roughly until there were further sharp declines during the 1930s. The reason for the decline was that the nation, despite its increasing population and economic development, required lesser amounts of wood. Per capita consumption of timber products was 156.9 cubic feet in 1900, and a steadily downward trend was already very evident. There was, in fact, no bottoming out, as per capita consumption passed below 100 cubic feet during the 1920s and continued to decrease.[19]

The timber industry was difficult to describe in generalities. It varied greatly from region to region, and it no longer was characterized

Forest conservation: natural regeneration of industrially owned Douglas-fir ten years after it was clearcut about 1940. The regrowth can be credited to care in harvesting, cleanup of debris after logging, and protection of the area from fire (courtesy Forest History Society).

by any particular type of economic organization or harvesting practice, although often the latter remained destructive. It was an industry burdened with taxable lands and timber, facing an uncertain future because of a declining demand for its products as the nation turned to masonry, metals, ceramics, and other materials for its buildings and implements. Nevertheless, many "forestry advocates" reached maturity when most large industries appeared to be monolithic, when monopolistic combines dominated the economy, and when arrogant logging operations wrecked many landscapes. Whatever its present nature or problems, the industry had a lot to answer for when it was brought before the bar of the Progressives.

The Progressive Era was associated with the presidential administrations of Theodore Roosevelt, William Howard Taft, and Woodrow Wilson. It was characterized by sweeping efforts to reform all sorts of social ills, and it resulted in the introduction of federal policies and programs into the lives of ordinary people. The Progressives themselves were politicians, journalists, reformers, and activists of every stripe, united in a general belief that the "little" people were at the mercy of the "interests" that had taken over the country. Among the objectives of the Progressive crusade were antimonopoly laws, pure-food-and-drug acts, the abolition of slums, the improvement of labor management and working conditions in the job place, and an end to child labor.[20]

The Progressives placed great stress on monopoly, which they viewed as the tool by which ''malefactors of great wealth'' were able to inflict a multitude of social evils on the people. Large concentrations of capital, the reformers charged, had brought America nothing but misery. The solution was to break up the trusts and to regulate the large industries so as to protect the good of the public.

The establishment of national forests fitted nicely into other Progressivist reforms. Along with the national forests, Gifford Pinchot and his confreres believed, must come federal control of industrial behavior on private lands, so as to end the ''cut out and get out'' tradition of the industry. By one means or another, industry must be made to serve the greater public good. However, even by the 1920s, federal control had not been extended to private lands. Foresters were left with only the tools of persuasion to institute sustained-yield timbering nationwide, persuasion made more difficult by the regulatory demands of strident reformers, ceaselessly castigating private enterprise.

The birth of forestry and the Progressive crusades left their mark on the Forest Service, reflected in its attitudes and its outlook. By the time the Service acquired the national forests, its foresters had accepted several things as articles of faith. One was that the nation needed wood and would need even more in the future. Another was the danger of a timber famine, which would be assured if destructive timber harvesting were to continue. The Service retained both beliefs tenaciously in the face of changes in the industry and of the nation's demonstrably declining dependence upon timber supplies.

Another characteristic of the Forest Service's origins was a technocratic outlook. Politicians and industrialists, it was believed, had long failed to serve the public interest because they were bound to their own desires for power and profit. Public affairs, however, had become too sophisticated to be governed by selfish ends. Rule by experts was the answer. Foresters, engineers, sanitarians, social workers, and a number of other educated elites were presumed to know best how technical questions ought to be answered; they were also presumed to be guided wholly by the light of reason and science. Many new federal agencies, especially those with regulatory duties or aspirations, were created during the Progressive Era, in the understanding that they would be run by trained experts who would serve the public good without personal preferment. Efficiency, not profit, would be their goal.

The new race of philosopher-kings proved to be somewhat less than they or others had hoped, of course. It was impossible to keep politics out of governmental administration in a democratic society; also, human nature was never simple, even if alloyed with a ''scientific'' education. The experts became progressively more narrow in outlook as a result of the kind of specialized education they encouraged. Pinchot and most of the leading Progressives were well-rounded men with broad interests. He was almost as concerned with the regulation of public power; the management of water resources; the prohibition of alcohol; the development of farm-to-market roads; the improvement of rural life; and other affairs as he was with forest conservation. The breed of young foresters that he helped to foster focused mostly on forestry, and from the perspective only of foresters. As long as the nation believed that foresters enjoyed a technical monopoly in forest management, there was relatively little harm in that. Trouble nevertheless lay ahead, because the foresters' view of the world was not the only one around.[21]

Pinchot's generation of foresters was the last for many years to disagree about the issues that most concerned it. Pinchot and the Progressive reformers believed that most problems could be attributed to the venality of big industry. Not all foresters, however, then saw the conservation issue as a simple question of public good versus industrial evil. Among those who took another view were Cary and Schenck, joined later by such people as William B. Greeley. They believed that industry simply acted in response to basic economic conditions. There was no incentive to conserve timber so long as it remained abundant and cheap. Schenck avowed that ''forest destruction'' must precede conservation. When timber became scarce and expensive, it would be treated as the valuable commodity it was. Schenck and the others predicted that conservation would not be adopted in the United States until it was economically necessary.[22]

THE DEBATE OVER CONSERVATION

Thus opened a national debate over the conservation of forest resources, with a view to developing a place for the national forests in the national timber economy. On the one side were the foresters, who for the moment appeared to be less interested in the national forests than in seizing technocratic control of the entire timber industry and its resources. On the other side was the forest industry, which had problems of its own. The foresters were not unanimous in their opinions about the nature and behavior of the industry or of the forces working on it. They appeared likely to become so, however, with Pinchot and the Progressives in firm control of the Forest Service and with the Forest Service—as employer of first resort—holding sway over the education of new foresters. Industry must speak for itself.

Early in the debate an industry spokesman told foresters that forestry "was impossible to apply" at that time. Outlining the economic realities of his business, he predicted that the pine forests of the North would have to be "sacrificed" before southern and western timber reached a value where sustained-yield harvesting would be possible. The need for profits would have to override all other considerations if there were to be production of timber. "When in the course of natural events," he concluded, "prices of stumpage have risen to the proper basis, other conditions being favorable, scientific forestry will surely be adopted by lumbermen."[23]

The question at hand was not the national forests; it was the fact that the majority of the nation's timber supply was on private lands. Foresters and timbermen faced the issue from different perspectives. The foresters' aims were the salvation of the forests—seeing forestry principles as the means to that end—and the avoidance of a "timber famine." The lumbermen wanted to safeguard their own industry; to that end they viewed forestry as either irrelevant or a threat. One lumberman said: "Some of the purely theoretical [forestry] may be wise, but the average mill owner has had considerable experience with the holder of theoretical ideas on this subject and looks with suspicion and disfavor upon the 'unhappy dreamers.' . . . [But the] greatest danger feared by the mill owner is that government action will not be uniform or accurately adjusted to the varying conditions of timber localities."[24]

The economic problems of the industry were real enough. Large blocks of prime timber were purchased or leased in the expectation of future returns. Operators built facilities and bought equipment with the same expectation. Most ventures were financed on credit, which intensified competition for available timber. The entire system depended upon a high volume of production and was risky enough without the unforeseeable effects of the harvesting principles advocated by foresters. Margins were so narrow that anything could threaten the stability of the industry, so even an operator sympathetic to the tenets of forestry could face failure if his competition kept to the old ways, or if times turned hard. "This competition during periods of commercial depression," the president of the National Lumber Manufacturers Association said in 1905, "might force the manufacturer who is practicing forestry to run his plant at a loss, or suspend operations until the conditions of supply and demand were favorable."[25]

Most of American industry would oppose federal regulation anyway, out of a natural belief in the principles of free enterprise. The argument for regulation of the timber industry, however, was based, more than anything else, on the purported threat of a timber famine. Timbermen, and others as well, were unimpressed with either the figures or the philosophy behind the foresters' eschatology. Industry leaders doubted that anyone could accurately estimate either the volume of standing timber or the rate of its depletion. Even Frederick Weyerhaeuser, one of the first major timbermen to experiment with conservative lumbering, thought the talk of timber shortages was exaggerated. "There is no reason to think," he remarked in 1909, that "the timber supply will not hold out indefinitely."[26]

Foresters and timbermen were in agreement on two subjects. One of them was forest fires. Both sides argued that it would do no good to practice conservation of resources if an investment in timber could go up in flames in a matter of hours. In promoting fire control, foresters and lumbermen were able to make common cause. The industry also received welcome support from foresters on the subject of tax reform. Both parties believed that prevalent systems of taxation threatened the security of the industry and prevented conservation, with investments eroded away in taxes if timber were left on the stump for future harvest. "Release the heavy burden of taxes on young forests not yielding immediate returns," argued Schenck; "save maturing forests from the short-sightedness of local tax assessors; protect young and old forests from fire and theft as well as any other property, and you will have forestry, because it will pay."[27]

Overriding all other considerations, in the foresters' view, was the need to get the nation to use its forest resources, public and private together, on a sustained-yield basis. So long as the foresters could not force their will on the industry by means of regulation, they must use persuasion. They had a long tradition of "forestry as propaganda," and the unsurpassed skills of Pinchot the propagandist to boot. Pinchot was from a business family, so he knew the way to the industry's heart: forestry can pay, he said. His colleagues echoed the same theme. "The fully equipped forester," Fernow wrote, "is a business manager, whose business it is to turn into profit the product of a forest property sustained in continuous revenue-producing capacity." Schenck also emphasized that "forestry on a large scale in the long run is not possible unless it be found to be remunerative one way or another." "The business of the forester," ventured the magazine *Forest and Stream,* "is to manage the forest so that the land owner and the lumberman can get out of it as much as possible."[28]

Pinchot started to carry the message to owners of private timber in 1898. He did it first with Circular 21, "Practical Assistance to Farmers, Lumbermen, and Others in Handling Forest Lands," which offered a way to introduce timber operators to the basic tenets of sound forestry practices. When invited by a company and at its expense, the Division of Forestry sent a team to inspect the private property and to develop a forestry program. The object was ''to provide successful examples of conservative lumbering.'' ''The main purpose of this working plan,'' said a typical example, ''is to outline a method of management under which the merchantable timber may be cut in such a manner that successive crops may be obtained and the condition of the forest constantly improved.''[29]

The Circular 21 program, in other words, aimed at sustained yield, or the balance between cutting and growth (''increment'')[30] of timber. The idea was to persuade the timber industry that sustained yield was synonymous with sustained profits. If the Circular 21 program were realized, the industry could stabilize itself over the long run by harvesting only mature trees, saving the rest to provide harvests and income in later years. Pinchot himself explained the principle:

> Conservative lumbering is distinguished from ordinary lumbering in three ways: First. The forest is treated as a working capital whose purpose is to produce successive crops. Second. With that purpose in view, a working plan is prepared and followed in harvesting the forest crops. Third. The work in the woods is carried on in such a way as to leave the standing trees and the young growth as nearly unharmed by the lumbering as possible.[31]

The Circular 21 program was moderately successful. In 1905 Pinchot reported that it had studied nearly 11 million acres of private lands (1.2 million belonged to one firm, Kirby Lumber Company of Texas). The program ended in 1909, however, under some questioning about the expenditure of public resources for private gain and in the face of the less-than-unanimous interest of the industry. Moreover, because the Forest Service had assumed charge of the national forests in 1905, it had less time available to attend to the lands of others.[32]

Before that happened, Pinchot and his associates tried to persuade the private sector that forestry did not mean a ''locking up'' of timber, nor even a loss of profits. ''Forest management and conservative lumbering are other names for practical forestry,'' Pinchot said, and ''practical forestry'' it often was called. ''Under whatever name it may be known,'' he emphasized, ''practical forestry means both the use and the preservation of the forest.'' A colleague added: ''The conservation thus suggested does not mean non-use of ripe timber, but does mean protecting it from useless waste and destruction, and replacing it by reforestation when it is used.''[33]

Many industrialists got the message. ''Practical forestry means conservative lumbering,'' said the president of a lumbermen's association. Accepting the definition was not necessarily to do things as Pinchot desired, however. By the end of World War I, he and others of like mind were not persuaded that the lumber industry as a whole had mended its ways. Things had changed since the great "cut out and get out" days of the previous century, but Pinchot and others reviled the entire industry. Regulation was the answer, they avowed. "Lumbermen as a class," scoffed one academic forester, "have failed to grasp the new conception of the relation between the ownership and exploitation of an essential natural resource and public welfare; and the adoption at some future time of a sane and practicable forest policy for handling the private forest resources of the United States must be done over the heads of lumbermen and not in co-operation with them, unless their attitude changes markedly in the near future."[34]

The Forest Service started a campaign of persuasion because, before 1905, its chief responsibility was public information. Pinchot turned that into a crusade for conservation on private lands, which, because his bureau lacked regulatory power, had to be one of propaganda only. The experience had lasting effects on the Service's collective character, reflecting Pinchot's messianic outlook and forestry's obsession with the nation's need for wood now and forever. The message was absolute, and the communication was mostly one way—the federal forestry experts preached to all but had difficulty in listening to others. In the short run, however, the political backlash against Pinchot's crusading caused the federal foresters to look inward. Pinchot's dismissal in 1910, during one of his outbursts, left the Forest Service feeling correct but bruised. Since 1905 the bureau had had on its hands the National Forest System, the greatest "demonstration" forest in history. There, it hoped, "practical forestry" could pass from propaganda to reality. Moreover, in a national climate that left technical programs to the experts, the Forest Service could develop its policies without discussing them with others—or so it hoped.

The Service received its marching orders from none other than Theodore Roosevelt, who said that the nation's forest policy was "not to preserve the forests because they are beautiful, though that is good in itself, nor because they are refuges for the wild creatures of the wilderness, though that, too, is good in itself; but the primary object of our forest policy . . . is the making of prosperous homes. . . . A forest which contributes nothing to the wealth, progress, or safety of the country is of no interest to the government and should be of little interest to the forester."[35]

Attitudes like that ensured that the national forests would be approached in essentially materialistic—or "utilitarian"—terms. That accorded with the materialistic outlook of the federal foresters, sure as they were that a growing nation must be founded on increasing supplies of wood and appropriately obsessed with the threat of a "timber famine." Convinced that it was right, fired with a sense of mission, and free of interference from others, the Forest Service addressed the national forests.

THE ROLE OF THE NATIONAL FORESTS

When the forest reserves were transferred to the care of the Forest Service in 1905, Gifford Pinchot sat down and wrote himself a letter (over the signature of Secretary of Agriculture James Wilson). It provided instructions for the fulfillment of the Service's new responsibility. The forest reserves were to be used, he said, because use was not contrary to conservation. In deciding what uses were to be permitted, the needs of local communities should come first. There was more: "In the administration of the Forest Reserves it must be clearly borne in mind that all land is to be devoted to the most productive use for the permanent good of the whole people. . . . Where conflicting interests must be reconciled, the question will always be decided from the standpoint of the greatest good of the greatest number in the long run."[36]

That letter was the Forest Service's Magna Carta; its philosophy and details permeate the agency's policies to this day. The evocation of "the greatest good of the greatest number" reflected the guiding spirit of the Progressive Era, in particular that of Pinchot, who was second only to Theodore Roosevelt as a spokesman—indeed a paragon—of the Progressives. He was on a crusade to transform American society, and the Forest Service was with him. As William B. Greeley remembered it, his people "had the thrill of building Utopia and were a bit starry-eyed over it."[37]

The national forests must play their part in a campaign to save America—save it from the timber famine, from the depredations of the "interests," and from the sorry social effects of America's way of harvesting timber. Seeking to build stable communities around the national forests, the Forest Service first must determine the social evils against which to guard its neighbors. It found three—communities ruined by "cut out and get out" lumbering, a homeless underclass created by the nature of the timber industry, and, above all, the threat to the timber economy allegedly posed by monopolies.

The Forest Service dispatched Samuel Trask Dana on an inspection tour of Pennsylvania and the Great Lakes states, to examine the social history of traditional lumbering. His report offered a number of instructive case studies. Cross Forks, Pennsylvania, for instance, had a population of five or six families when a lumbermill opened there in 1894. The community grew quickly to about two thousand people, with seven hotels, four churches, a railroad, a YMCA, a waterworks, an electrical plant, and other symbols of stability. The timber had been consumed within fifteen years, however; the mill had closed, and the scene was one of desolation. Fires were so frequent that the insurance companies had canceled all policies. Houses were on the market for $25 to $35 but found no buyers. Cross Forks was a ghost town.

All around the Great Lakes there were other ''deserted villages,'' surrounded by barren areas that had once been forest but now were ravaged by great fires that consumed brush and logging debris. Railroads had been abandoned, and timber, once so plentiful, had to be imported from afar and at high cost. ''The moral,'' Dana recalled years later, ''was that the way to avoid ghost towns was to practice sustained yield forestry.''[38]

Dana also took note of working conditions in the timber industry. What he found was frightening—exactly the kind of rootless, migratory underclass that Progressives regarded both as a national disgrace and as a potential menace to publc order. Lumberjacks could not lead an ''ordered life,'' Dana thought, and a lumber camp was not ''the best place to raise a family.''[39]

The Forest Service sponsored a study of working conditions in the timber industry, under the authorship of Benton MacKaye. He reported the findings of a presidential commission that 90 percent of the work force of the timber industry comprised unmarried men, with a turnover rate of 600 percent. '' 'Timber mining,' '' as MacKaye and others called traditional practices, ''being itself a tramp industry, is a breeder of tramps; it is an industry of homeless men.'' In contrast, he believed, putting forest properties on a sustained-yield basis would allow permanent communities of family men to replace the migratory camps. In addition, such reform of timbering would create seven hundred thousand permanent jobs, which would support a stable population of three and one-half million people.[40]

Progressive conservationists were quick to point fingers at the cause of such evils as devastated communities and hobo workers: it was the concentration of wealth and power embodied in economic monopolies. Pinchot wrote: ''Monopoly on the loose is a source of many of the economic, political, and social evils which afflict the sons of men. Its abolition or regulation is an inseparable part of the Conservation policy.'' Accordingly, his first *Use Book* for the national forests (1905) promised, in regard to the sale of timber, ''There is no limit . . . to the quantity which may be sold to one purchaser, but monopoly to the disadvantage of other deserving applicants will not be tolerated.'' Two years later, in the revised edition, he observed, ''There is no chance for monopoly [when] the local demand is always considered first.''[41]

The national timber industry was actually too diverse to be dominated by trusts or monopolies after the turn of the century. Even the larger enterprises failed to control the business of their regions entirely, although some exerted great influence, as was the case on the Pacific Coast, where considerable concentration of ownership took place early in the twentieth century. Nevertheless, the Forest Service believed that when it came to managing the national forests, monopolies were a real menace, not merely Salem witches. Its examination of the lumber business in California in 1908 concluded: "Many of the larger lumber companies have combined into a trust," led by Valley Lumber Company, which allegedly controlled retail prices throughout the San Joaquin and Sacramento valleys. "Until the privately owned stumpage is exhausted," timber sales officers were warned, "it would be foolishness to throw Government timber into the market controlled by the lumber trust."[42]

Austin Cary made another investigation of the lumber business in the San Joaquin Valley in 1913; he offered similar findings. Reviewing dealers' sales lists, he concluded that "lumber is sold to the people on a scale of prices that can only be the result of agreement." Furthermore, he found evidence that dealers divided the market among themselves. Worst of all, he detected collusion among parties buying national forest timber, thus lowering the prices received by the government. "Purchasers in the San Joaquin Valley, in my opinion," he ventured, "have thought that by action concerted among them they could get more or less timber of the Forest Service at a price of their own naming."[43]

Cary was not alone in his suspicions, nor was the threat that monopolies allegedly posed to the national forest's timber program isolated in California. The Bureau of Corporations warned, at about the time of Cary's investigation in California, that the timber of the entire nation was within the grasp of a gigantic monopoly, and worse. "There has been created not only the framework of an enormous timber monopoly," the bureau claimed, "but also an equally sinister land concentration." Fear of monopoly, real or not, was the Progressive order of the day.[44]

The message was clear for the Forest Service. If it wished to protect itself from charges of collusion with industry, ensure that timber would be available for the home builder, and receive a fair return to the government, it would have to use the greatest of care in devising policies with regard to the sale of timber. In guarding the national forests from the presumed evil of monopolies, the Service sought also to protect communities that would become increasingly dependent upon those forests.

In determining to manage the national forests as instruments of social reform—promote community stability, institute sustained-yield harvesting to stave off a timber famine, improve the lot of the lumberjack, fight monopolies—the Forest Service took on a tall order. It also acquired a wide range of opponents, some of whom were averse to its policies; others, to its philosophies; yet others, to the whole conservation idea; and some, simply enraged by the barbs of the wide-ranging, ever-combative Pinchot. Worst, perhaps, was the hostility of much of the West to the very idea of forest reserves, virtually from their beginning in the 1890s. Graziers, the most prevalent land users, feared restrictions on the use of range and water. Prospectors feared the loss of mining opportunities. Water rights—always a sensitive issue in the dry West—were a concern everywhere. And speculators of every stripe naturally objected to the imposition of rules and controls that had previously been lacking.

The Forest Service saw itself in the midst of competing interests, itself as the proper arbiter of the public good. All questions came under the test of the "greatest good of the greatest number in the long run." The long-term interests of communities served by national forests, the foresters believed, could not be promoted with actions that would be beneficial only in the short term. Timber shortages threatened—always. Communities that could rely on sustained-yield timber policies would prosper for centuries. In that frame of mind, therefore, the Forest Service set about winning the West, by persuading communities that national forests were valuable assets, not eastern-bred restrictions on their growth.[45]

The need to win western support reinforced the foresters' materialism. The highly materialistic Pinchot, ever the publicist, began by proclaiming the good news from Washington immediately after he received charge of the forest reserves. The Department of the Interior, following the lead of the 1897 legislation, had acknowledged but two purposes of the reserves. They were "to furnish timber" and "to regulate the flow of water." According to Pinchot, they should be "for the purpose of preserving a perpetual supply of *timber for home industries,* preventing destruction of the forest cover which regulates the flow of streams, and *protecting local residents from unfair competition.*" Two years later, in 1907, he promised that the 1897 legislation "made it possible to use all the resources"—wood, water, minerals, forage, and recreational values—of the national forests.[46]

Of all the resources, timber received particular attention in the campaign to sell the national forests to the western public. The Organic Act of 1897 had accorded the "free use" of timber to "bona fide settlers, miners, residents, and prospectors for minerals, for firewood, fencing, buildings, mining, prospecting and other domestic purposes.'' That measure, Pinchot believed, ''made friends for the Forest Reserves and helped the settlers and the little men whom we wanted most to help.''[47]

''We know that the welfare of every community,'' the Forest Service assured its neighbors in 1905, ''is dependent upon a cheap and plentiful supply of timber.'' Two years later it said: ''The fact should never be lost sight of that [the national forests] are for the homebuilder first, and that their resources are protected and used for his special welfare before anything else.''[48]

Pinchot's campaign of persuasion received a significant boost from Congress in 1906. The Agricultural Appropriations Act that year directed the return of 10 percent of proceeds from the sale of timber on the national forests to the counties in which the forests were located. Two years later the percentage increased to 25. Those measures not only countered arguments that reservation of the forests would deprive counties of future property taxes; they made the sale of national forest timber a matter of immediate economic importance to the affected communities.[49]

Propaganda and distributed funds were not enough to persuade communities that national forests were at their service, however. Pinchot must establish an organization to carry the message directly to the people. When the General Land Office administered the forest reserves, decisions affecting them emanated from Washington. Aggravated by a shortage of personnel, that system led to delays, mismanagement, and public frustration. Even before he had charge of the reserves, Pinchot decided ''that in the matters of judgment concerning forest problems, the field man who knows the conditions, should rule.'' Because of his devotion to the propaganda campaign and his other political commitments, Pinchot delayed establishing a new administrative structure, finally acting when prodded by Secretary Wilson. By 1908 the Forest Service had been reorganized and the administration of the national forests decentralized.[50]

Pinchot established six geographical Districts (now called Regions), with headquarters in Missoula, Montana (D-1); Denver, Colorado (D-2); Albuquerque, New Mexico (D-3); Ogden, Utah (D-4); San Francisco, California (D-5); and Portland, Oregon (D-6). Within the Districts, a forest supervisor had charge of each national forest, which was subdivided as appropriate into ranger districts. Authority was delegated downward insofar as possible; even district rangers made important decisions without first getting permission from above. That had a decidedly positive effect on relations between the Forest Service and local communities. Citizens could receive an immediate response to their advances, and rangers who were answerable to their neighbors

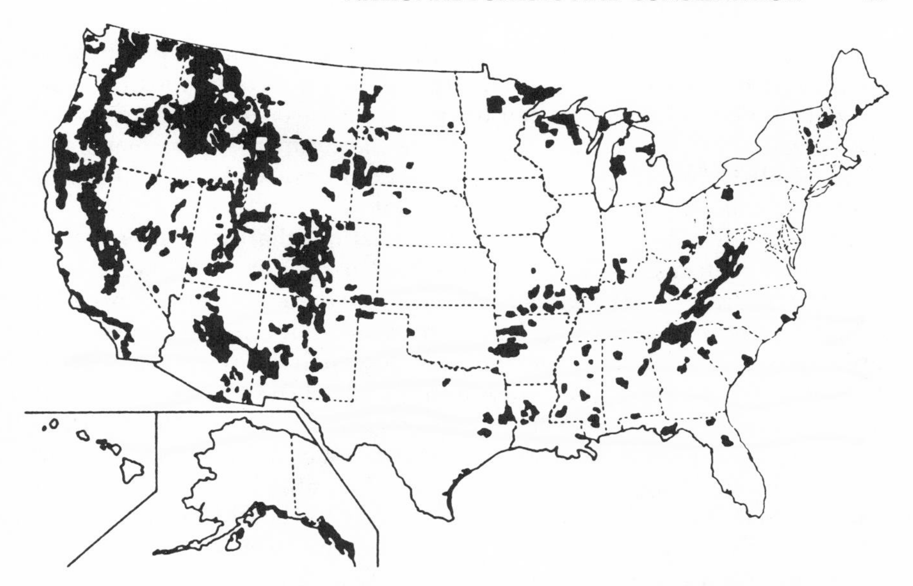

The National Forest System in 1980. In the West there has been little change since 1907. In the East forests were purchased under the Weeks Act of 1911.

behaved differently from those who could pass the buck to an office a continent away. "Handle your districts and take as much interest in your work as if they belonged to you," one supervisor told his rangers. However, he reminded them, never forget whom you work for. "Think Forest Service during the day and dream it at night and you will come through in good shape."[51]

That sense of being a part of something important, that esprit de corps, was the most important of the many enduring legacies that Pinchot gave to his agency. The Forest Service had a special mission, he maintained, and his people and their successors agreed with him. That mission was to be of service to all the people, and especially to the communities that depended upon the national forests. In pursuing that calling, the Forest Service looked first to timber.

"Communities will depend upon the National Forests for a steady supply of timber," said George Cecil of District 6 at Portland, "and if we cannot meet this demand, we shall have failed in our mission. . . . [It is] doubly important that we regulate National Forest cuttings with the greatest consideration for the future welfare of the local communities."[52]

Elsewhere, in District 1, which encompassed the northern Rocky Mountains, it was the unanimous observation of the forest supervisors in 1910 that ''sales are particularly desirable in the development and maintenance of local industry.'' To that end, local companies that did not have timber holdings of their own ''should be encouraged'' by sales of national-forest timber, thereby making the companies ''permanent.'' Over all other considerations, the supervisors avowed, ''the needs of the community [for timber] are paramount.''[53]

Pinchot left behind him an organization that was thoroughly dominated by foresters with an outlook all their own. They were on a righteous crusade to guarantee more wood for the nation and to prevent a timber famine. They must demonstrate the value of sustained-yield timber management by showing how national forests could be enduringly productive for the communities around them. They must also persuade the much larger private timber economy to follow their example. Moreover, they were the experts who alone knew how the nation's timber ought to be managed—''scientific'' gentlemen who believed that their principles were wholly technical and free from self-interest. They were accordingly inclined to keep their own counsel. In managing the national forests, foresters would talk mostly to other foresters, to ensure that in their view the job was done right. Reforming the rest of the timber community would follow in due course.

The Forest Service had promised to take over the national forests and to manage them for ''the greatest good of the greatest number in the long run.'' For a forestry service, that had to mean, above all, the management of timber. The organization and its fundamental doctrine were firmly in place. Now it was time to translate propaganda into practice.

CHAPTER TWO

Forging the Timber-Management Program

In one sense, there are no purely unselfish motives touching economical questions. Even the disciples of forestry wish to preserve, propagate, and perpetuate the forests because the results would be of economic benefit to mankind.

—J. E. Defebaugh (1905)

The Organic Act of 1897 authorized the sale of timber on the forest reserves. The guidance it offered in making sales, however, was limited: "For the purpose of preserving the living and growing timber and promoting the younger growth on forest reservations, the Secretary of the Interior, under rules as he shall prescribe, may cause to be designated and appraised so much of the dead, matured, or large growth of trees found upon such forest reservations as may be compatible with the utilization of the forests thereon and may sell the same for not less than the appraised value in such quantities to each purchaser as he shall prescribe."[1]

In implementing the terms of the 1897 legislation, the Forest Service must step down from the stage of principles and theories into the real-world arena of timber sales. It spent nearly a decade grappling with that issue, torn between political insistence that it become self-supporting through sales receipts and the indifference of the timber economy, expressed in a low demand for national-forest timber. Since the Service began in considerable ignorance, it therefore developed a research program as a basis for its sales program. In addition, there were many other public demands on the forests besides timber; because they competed for the attention of a timber-minded bureaucracy, they also must be acknowledged. Most important was the thorny problem of appraisals before sales, to ensure valuations that would be satisfactory to buyers and sellers alike. The Service solved that problem wholly on its own, and when it was through, it had propounded an appraisal system that endured almost unchanged thereafter. The federal foresters' isolation from the affected publics in that exercise was harmless enough at the time, for the volume of sales was minuscule. Nevertheless, the experience set the stage for later controversy, because it left the Service inwardly focused and hardened in its attitudes, devoted to a concept of sustained-yield forestry founded on the belief in an impending timber famine that to everyone else seemed progressively less likely.

SELLING THE TIMBER

The timber-sales authority granted to the secretary of the Interior in 1897 was delegated to the commissioner of the General Land Office. Commissioner Binger Hermann's *Rules and Regulations Governing Forest Reserves*, issued in 1897, required each applicant for federal timber to petition the local land office. The local officer would examine and appraise the timber and then send the petition to Washington, D.C., for a decision on the merits of the request and the appraisal.[2]

The prospective buyer's application included a price offer. If his bid accorded with the forest officer's analysis of local timber prices, the petition would go to Washington with the officer's recommendation. "I believe," one supervisor typically attested, "the price offered for this timber is higher than that paid to private parties for like timber in the same locality."[3] Other than trusting the officer's word, Hermann had no way of judging the relative merits of each application. The 1902 version of the General Land Office's *Forest Reserve Manual* simplified the rules. It said very little about appraisals, however, and in fact made no attempt to impose an appraisal system.[4]

The reserves were transferred to the Department of Agriculture in 1905. Very few procedural changes occurred until decentralization in 1908. The first *Use Book* (1905) followed the basic principles established by the General Land Office. Stumpage prices, forest officers were told, were of "great importance" and "should be decided not by custom or habit, but by actual value of the timber as determined by its character, ease of logging, and distance from market." The book offered no appraisal formulas, and the Forest Service reserved the approval of most sales to the Washington office.[5]

The failure of the General Land Office and the Forest Service to formulate appraisal procedures was less a matter of intent than of priorities. Few sales were made in the early years, and timber prices seldom exceeded a dollar or two per thousand board feet (MBF); the government sold only 68 million board feet (MMBF) in Fiscal 1905, valued at $86,000. Inadequate manpower faced more pressing concerns, while timber receipts did not justify expensive studies. A little bargaining over price seemed a reasonable alternative. "In fact," California timber sale personnel learned, "there is little more actual basis of valuation in selling timber than in selling horses."[6]

The Washington office began to address standardized stumpage rates in 1908, when sales reached 525 MMBF, at $964,000. Associate Forester Overton Price canvassed the Districts on the advisability of adopting "flat rates," as opposed to individual cost studies on each timber sale. The aim, he said, was to "prevent the danger of selling timber in certain forests at a lower rate than timber of equal value on other Forests in the same region."[7]

Each District initiated a study of the recommendations in Price's memorandum. District 5 established minimum rates for each of the California national forests. Those in which there was competition for sales received rates for both live and dead timber; elsewhere only one price prevailed. "Other conditions being the same," the District's chief of silviculture explained, "it is very desirable that two or more Forests supplying timber to the same market or class of markets should have the same stumpage rate for like material in order that no injustice may be done purchasers."[8]

The District office in Missoula asked Chief Inspector Edward A. Sherman for recommendations. Instead of minimum rates, Sherman suggested "standard" and "minimum" prices for each forest, the former for the "general market," and the latter for "small sales where the material is used locally." The two rates, termed "maximum" and "minimum" by the Washington office, were accepted in District 1 and became the "established rates" for all forests in the northern Rockies.[9]

The application of maximum and minimum rate structures soon aroused complaints from the Districts.[10] The Washington office told them in March 1910 that problems would be avoided if supervisors were not "tempted to secure sales from each other by shaving prices." The Districts therefore should adopt a "schedule of standard prices," based on the "present minimum stumpage."[11] Dissatisfaction over that suggestion soon caused a reevaluation of appraisal procedures.

Monopoly, speculation, and allowable cutting levels also delayed the adoption of appraisal methods by the General Land Office and by the Forest Service. The need for timber sales, both to obtain working capital and to place the national forests under intensive management, had to be balanced against fears of abuses that might result if policies were adopted without considering their impact upon the long-range goals of federal forestry. Sales and appraisals made in haste could work contrary to that duty. If the volume of timber sold to any one purchaser precluded sales to others, the effect would be monopoly. If appraisals were too low, bidders would resist cutting timber, in the expectation of rising prices. If long-term contracts were adopted without scheduled harvests, industry might cut at irregular intervals, to the disadvantage of forest management and community welfare. If appraisal methods overpriced the timber or if stringent cutting controls were imposed, there would be no buyers. Such concerns bedeviled those who were trying to develop a timber sales program.

Binger Hermann imposed the restriction that timber must be removed within one year of the date of sale. If annual consumption exceeded one million board feet, an extension could be granted, but prospective buyers must apply six months in advance, so that cutting could begin and end within the one-year limit. Even then, applicants could not be certain of winning an entire award. The commissioner could "make allotments of quantity to several bidders at a fixed price, if he deems proper, so as to avoid monopoly."[12]

Those terms worked to the disadvantage of the timber industry and the sales program alike. Investors were reluctant to spend money on plant and equipment when full-volume awards were not assured or if the possibility existed that the investment could not be amortized over a long-enough time. Development costs required contract terms that would provide either more volume over extended periods or the assurance that subsequent sales could be expected.

The Forest Service chose to pursue long-term sales after taking charge of the national forests. Sufficient volume for a five-year harvest was offered, along with the proviso that "definite" quantities must be removed each year. The foresters believed that "in this way, speculation in National Forest timber is prevented." Signs of caution appeared in 1910, when the Washington office ordered the Forest Service to "split sales if necessary to prevent monopoly."[13]

Five-year sales allayed some industry concerns, but they also fostered speculation. Owners of private timber took advantage of unstable retail prices for lumber by holding their timber during market downturns and bidding for federal timber at the lower prices. Sales of national-forest timber fluctuated in consequence, from 525 MMBF in 1908 to 458 MMBF in 1909 to 484 MMBF in 1910, while the total national production of saw logs remained roughly the same. Moving in and out of the national forests, operators kept their plants and equipment in operation, secured capital, and gambled that future markets would improve enough to allow a profitable return to their own timber.[14]

Many foresters thought such practices were inimical to conservation. Pressure to sell more federal timber threatened sustained harvests and, when higher prices returned, encouraged owners to overcut their own lands. Accordingly, William Cox said in 1910 that "the Service will not sell to companies holding large areas [of timber] for speculation."[15]

That policy received a cool reception in some Districts. The Missoula office told its supervisors that the policy was a general statement of intent for the "whole West" and therefore did not necessarily apply

Henry S. Graves, second chief of the Forest Service. He healed the Service's wounds after Pinchot was dismissed, and he oversaw the completion of timber-sales policies, including the appraisal manual (courtesy Forest History Society).

to "local conditions." "The policy pursued in this District must of necessity be determined by local conditions, thus making it necessary in some cases to vary more or less from the general policy laid down in the [Cox] memorandum."[16]

The Washington office pounced on that. Forester Graves asked District 1 to be more specific. Had Missoula intended to apply its principles to the District's silvicultural policy or to the Washington office's policy in general? Graves wanted to know which parts of Cox's statement "may not be applicable, in order that we may get it absolutely right."[17]

William B. Greeley replied that denying sales to firms holding private timber would not be in the best silvicultural interests of Missoula, because it would prohibit transactions with the Anaconda Copper Mining Company and with the Big Blackfeet Milling Company. To forgo sales to those two firms would virtually shut down silvicultural activity on national forests near them.[18]

Neither District 1 nor the Washington office argued the point further, and sales to firms holding their own timber continued throughout the West after 1910. Worries about speculation were overshadowed by the projections of a foreseeable timber famine. Private holdings eventually would reach critical levels, and the federal foresters, according to T. D. Woodbury, would "be in a position to impose conditions upon purchasers which tend toward good forestry, the enforcement of which today is impracticable."[19]

Before 1910 the producing capacity of a forest was a matter of guesswork. For that reason the Forest Service adopted allowable cutting levels, to guard against overcutting. "Allowable cut" denoted the amount of timber removal that would produce sustained-yield harvests. Theoretically, allowable cut was determined by finding the maximum volume that could be harvested periodically (usually annually) in perpetuity. That required a knowledge of the total volume of timber on hand, the age at which a species reached its maximum growth, and the typical rate of replacement of a harvested tree by a seedling. Aware that it would require considerable research before reliable estimates would be possible, the Washington office determined that the allowable cut should reflect sales during fiscal year 1907. In California, for example, the total stand of one national forest was estimated at 8 billion board feet (8 billion BF). Dividing that by the 1907 harvest of 40 million board feet (MMBF), two hundred years would elapse before each stand of timber would have been cut once (one "rotation"). District 5 said that a more accurate annual allowable cut could be obtained by multiplying the annual net growth of each acre by the total acreage. If growth amounted to 300 BF per acre on 450,000 acres, District 5 should be able to increase its annual allowable cut to 135 MMBF. The Washington office compromised between its own figures and the District's, allowing a harvest of about 1 percent of the estimated live timber (one-hundred-year rotation) every year.[20]

A. B. Recknagle thus explained the principle of the annual allowable cut to the forestry profession and the public in 1915:

> Every Supervisor is allotted a certain annual cut, based on the best estimates attainable. This annual cut he treats like a bank account. The limitation for each Forest is approved by the Secretary of Agriculture. Of course, it is not expected that the Supervisor will necessarily use up the limitation each year, so he can either let it accumulate for a number of years, or else he can make a larger sale than the limitation allows, prorating it for several years in the future. This limitation serves the purpose of an effective "lid" on over-cutting.[21]

Coincident with the discussions on allowable cut was a significant decline in some retail lumber markets, beginning in 1910; total national production of all forest products fell from $3.5 billion to $3.1 billion from 1910 to 1915, with wide fluctuations on the way. Those who believed that the government should refrain from allowing sales during periods of depression in the industry argued against compounding the problem. If sales were to continue, William B. Greeley opined, it would be preferable to adopt an "over-cutting" policy for species unaffected by the recession and to leave the affected species until the trend had reversed. The roles then could be switched. All species, concluded Greeley, would then "contribute their proportionate share of revenue."[22]

Austin Cary agreed, writing: "My dear Greeley, With you I have felt [heretofore] that timber sales might be easily pushed too hard for the permanent good of the Service. Rapid disposal of the timber in the National Forests is not what people at large are expecting of us." He suggested that a surplus of old-growth timber existed, but he was not ready to concede that intensive management was therefore necessary. Centuries of time had created the surplus, and even should a timber famine occur, Cary did not believe it would be possible to replace the entire volume with young growth and thereby have the stands at full producing potential in a reasonable time. He thought the "*reservation* idea" should be reinstituted for the national forests.[23]

In contrast, Burt P. Kirkland outlined opinions that were held more generally in the Forest Service. Dependent communities, he said, would lose jobs, as well as 25 percent of the federal timber receipts during any harvesting moratorium. More important, he stressed that by placing national forests on a sustained-yield footing at the earliest possible date, the Forest Service could minimize the effects when private timber could no longer meet the nation's needs.[24]

Kirkland's defense of immediate harvests reflected the long-range silvicultural objects of federal policy, which was based on an interpretation of forests that would characterize federal foresters throughout the twentieth century. That view held that old-growth ("overmature") timber was unproductive, because it did not grow fast. It should be replaced with younger trees, which would increase the total volume of wood in the forest at a faster rate. Volume was almost the only consideration; the quality of the wood was secondary, and other values of the forest were not addressed. If the national forests were to hold only "overmature" timber when the "timber famine" struck, they could only stave off disaster until they also had been consumed. If, however, the national forests contained mostly fast-growing trees that were approaching maturity, sustained yield would see the nation through.

Kirkland's statement of purpose did not betray the real reason for increased timber sales, however. Congress expected the federal forests to become self-supporting. As each year passed, the Forest Service asked for funds beyond those received from timber and grazing fees. By 1911 the Washington office had grown sensitive to criticism of its inability to balance income and expenses. Henry Graves wanted to end the annual budgetary confrontation. In 1911 he called for a "special effort" to reach "self-supporting" levels on each national forest. Three years later he called the budget problem "urgent and critical." Every year, he complained to the district foresters, "I am forced to meet this insistent criticism [which] cannot be ignored."[25]

Graves believed that despite a slumping lumber market, each District could increase its income by aggressively pushing local sales and by employing the idea of "substituting a cut of Government timber for a corresponding cut of private timber." In other words, private timber holders should be encouraged to refrain from cutting on their own lands in return for federal stumpage—a flat reversal of earlier antispeculation policies. Graves argued for careful appraisals, to encourage sales and check monopoly. Regarding the latter, he explained: "The Forest Service could not commit suicide more quickly or more effectively than to permit the impression to get established that it is seeking to aid private owners of timber in restricting competition in the production of forest products." Stumpage, he added, should not be sold at any "material sacrifice" of market value. Appraisals must be "liberal rather than conservative," Graves concluded; otherwise the Forest Service would have no way of justifying a lack of future sales before Congress.[26]

Forest Service attitudes toward the volume of timber that could be offered in single sales began to change. The possibility of offering larger volumes of timber, over longer periods of time, seemed attractive when compared to the limited outlets afforded by local mills and local markets; vast areas of overripe timber, remotely located in underpopulated areas, required an imaginative policy. If big capital could be induced to invest in such ventures, the Forest Service would reach its objective of increased sales volume, bringing the national forests under intensive management quickly.

For five years the Forest Service had publicized the relationship between the national forests and dependent communities. A particularly vivid promise exemplified the Service's vision of the future: "Scattered everywhere through the mountain land will be many small mills . . . to furnish the men of small communities work through the winter months, and the bulwark of manly and sterling independence." However, an advertised offer in 1910, to sell at one stroke a billion board feet of timber as a favor to a railroad, belied both that bucolic picture of national forest communities and the Forest Service's directive that "sales of small amounts of timber are preferred and will be encouraged by every means possible."[27]

The contradiction in policy was not as sharp as it appeared. After five years of administering the national forests, the Forest Service realized that local markets alone would not be able to exploit the vast resources of the federal timberlands. The Districts therefore began to differentiate between "strictly local markets" and the "general competitive markets" as the volume of local demand became more apparent. In District 4, for example, 95 percent of the sales went into local markets. That represented only 20 percent of the District's federal timber, making 80 percent of the allowable cut available for the general market. Upon such assumptions and always with the caution that "small operators are essential to the welfare of local communities needing lumber . . . and that it is the policy to encourage small operators," the Forest Service moved toward long-term, large-volume sales.[28]

Authority for large sales was officially recommended to the secretary of agriculture in 1911. Citing the pressure on the Forest Service to become "self-supporting," Graves aimed at sales of up to 500 MMBF, over a ten-year period, with the stipulation that sales should be restricted to areas where a "heavy initial investment was required." Graves warned that "the situation of the applicant should also be considered."[29] Apart from "considerations of local policy"—which Graves defined as the encouragement of small operators who would supply local industries, as against large operators who would export timber and "endanger the local supply"—the determination of whom timber should be sold to "should depend upon which course would be most advantageous to the Government." Graves added that "if local market and logging conditions are such that exploitation of the timber by small operators will net the best returns for the stumpage, this plan should be followed; if large operations, with their more efficient plants and lower unit cost of production will bring the better stumpage price, the timber should be held for sale on this basis."[30]

Minor objections were raised about the profitability of increasing timber sales in a depressed retail market—particularly in District 5, where it was believed advisable to delay sales until the opening of the Panama Canal improved California markets and, accordingly, Forest Service prices.[31] The majority of federal foresters, however, favored the silvicultural benefits expected to come out of the new policy. The only major question was whether five-year sales would provide a sufficient inducement to investors.

District 6's Forester George Cecil in Portland believed "that twenty year contracts should become the rule rather than the exception." He based his conclusion on the "business foundation" that is required in order to attract capital and on good silvicultural policy.[32] Two District 6 foresters questioned making long-term sales in the absence of an adequate timber-appraisal system. They queried Washington in 1912 about the possibility of continuing the policy of limited contracts, with the inducement of guaranteed future sales to the original purchaser. The Washington office disapproved: "The most we can do is to agree to advertise further blocks of timber in the unit when the first [short—term] contract is completed."[33]

On 6 February 1913, Graves modified the concept as it applied to short-term contracts: "Where the exploitation of National Forest timber requires new improvements, I believe that we should reserve enough timber to give these improvements a reasonable operating life judged by ordinary commercial standards, although only a portion of such timber may be included in the initial sale." He also cautioned that this "does not mean that the buyer of the first [sale] block is guaranteed the purchase of the second or other subsequent blocks." The Forest Service would only "withhold this additional timber from sale until the initial contracts are cut out," at which time the established firm would have to bid against any other potential buyers. The fear of monopoly precluded any other course.[34]

The pressure to increase timber sales led sales officer Franklin H. Smith to suggest a plan for marketing District 1's surplus timber. The "chances," as Smith termed them, should be advertised for the benefit of potential investors in markets in which the awareness of Forest Service timber sales was not common. Through "some central organization," the advertisements could be "followed up systematically and persistently" with interested buyers. District Forester Ferdinand Silcox forwarded the proposal to Graves in October 1911, suggesting that Smith get the job of implementing it. Graves approved it in November.[35]

Smith served as the sales liaison between the Districts and potential buyers until 1915. The Districts forwarded information on high-volume sales to his Chicago office for review, publication, and follow-up. A typical sales prospectus offered to investors the information they needed, discussing the volume of the stand, the terms of the sale, the principal markets, transportation facilities, and special requirements. Advertisements in trade journals and periodicals alerted the lumber industry and the financial community.[36]

While the Forest Service pursued its timber-sale and appraisal responsibilities during its first decade, it seldom solicited the views of outsiders. It did keep an eye on silvicultural practices, however. The Service simultaneously started to work on the study of tree growth, tree measurement, reforestation, cutting practices, and site classification. In many respects, the administration of timber sales could not proceed farther until research had produced information to support it.

ORGANIZATION AND COMPETING USES

Congressional intent to manage the forest reserves according to the best principles was inherent in the enabling legislation. The statements "to improve and protect the forest within its boundaries" and "to furnish a continuous supply of timber" constituted a mandate for the emerging discipline of forestry. Foresters transformed the actions required by the legislation—"improve, protect, and furnish"—into their creed of "sustained yield," or the process of supplying regular quantities of timber in perpetuity. As defined by the Forest Service's early *Use Book,* sustained yield meant that "cuttings do not damage the Forest, because the lumbering operations are so carefully done that the stand is left in first-class condition for a second crop, and after that a third crop, and any number of future crops." In Pinchot's view, timber management in the Forest Service was integral to the existence of the forestry profession: "The business of foresters is to manage forests, as the business of farmers is to manage farms."[37]

Timber management was originally assigned to the Branch of Forest Management, which was subdivided into departments of Timber Sales, Cooperation, and Inspection, which included Silviculture and Logging sections. The Branch of Publication and Education also had an Office of Silvics, with separate departments addressing field investigations; while the Branch of Forest Extension managed Reconnaissance and Planting Stations under the Office of Reserve Planting.[38]

In 1907, timber activities were consolidated under the Branch of Silviculture. Three offices—Management, Silvics, and Extension—supervised the various duties associated with forest management. The chief of the Office of Management directed cooperation, timber sales, reconnaissance, and an atlas of federal timberlands. His six chief inspectors toured the Districts to "see exactly" what was being done. The Office of Silvics studied forest and tree distribution, stream flow, climate, and regeneration. The Office of Extension supervised forest planting and advised private landowners. Those divisions of authority continued until 1920, when Silviculture again became the Branch of Forest Management. In 1937 a new title, Division of Timber Management, appeared. The organization is now called the Timber Management Staff.[39]

In addition to timber, the Forest Service acknowledged other responsibilities. Watershed, recreation, grazing, and mining had been major expectations of people who advocated a federal forest policy throughout the debates on the legislation. The federal foresters therefore sought allies across the broad spectrum of conservation interests. Any function that was associated with the forests—be it fish and game, hydroelectric power, navigation, tourism, irrigation, fire control, recreation, livestock grazing, or water supplies—was trumpeted in the interest of forestry. Very few forestry publications lacked rousing affirmations of the benefits to everyone if sound forestry programs were adopted. "Why Anglers Should Become Members of the American Forestry Association" was a typical argument reflecting the breadth of the campaign to win converts to the cause of the "multiple use" of national forests.[40]

Multiple use in the Forest Service evolved from legislative intent and out of a need to broaden the support for forestry. It first took root in the General Land Office, which had the multiple responsibilities of developing policies for timber sales, water policy, and grazing permits; which issued special permits for "public stopping places, hotels, stores, etc."; and whose officers acted as "assistant[s] to the game wardens of the State or Territory in which [they are] located." It was a difficult notion to define, let alone to administer. In 1907 the Forest Service offered broad guidelines for the management of lands that were subjected to conflicting demands: "There are many great interests in the National Forests which conflict a little. They must all be made to fit into one another so that the machine runs smoothly as a whole. It is often necessary for one man to give away a little here, another a little there. But by giving a little at present they both profit by it a great deal in the end."[41]

Another half-century would pass before the concept of multiple use received much attention outside the Forest Service, however. Concern over timber famine lay behind the creation of the national forests and the emergence of a forestry profession. Most foresters concentrated on the methods needed to increase timber supplies, instead of on the multiple use of the forests. Achieving sustained timber harvests, accordingly, outweighed other responsibilities. Even in that, however, achievements were modest by later standards, as national-forest management remained "custodial" for many years. Commercial sales of national-forest timber did not exceed a billion board feet until 1924, and then they represented a tiny part of the national timber economy, which that year produced nearly 100 billion board feet.[42]

TIMBER MEASUREMENT AND WORKING PLANS

The rate at which timber supplies were being depleted—depletion was a basic assumption of foresters—defied description. According to one Forest Service study, annual consumption in the United States was "four times as great as the annual increment [growth] of our forests." The federal foresters believed that in the absence of reforestation, no one could reliably estimate the long-term effect of demand. That heightened their resolve to achieve sustained yield. In fact, however, they tended to pick through the figures to see what they expected to see. The national consumption of industrial timber products totaled about 103 billion board feet in 1908. It increased slightly the next year, then held roughly level until a decline set in in 1914. Consumption was 80 billion board feet in 1921, just before a slight rebound to around 100 billion board feet; then there was a renewed decline during the depression. The Forest Service persistently ignored those trends—despite the fact that it compiled the statistics. The timber famine, which seemed ever more unlikely to the nation, remained real to the foresters.[43]

Given the Forest Service's outlook, the institution of sustained yield seemed immediately essential. The first step was to ascertain, with some precision, the extent of existing timber stock. General Land Office and United States Geological Survey mapping of the forest reserves had begun, but the volume and species of available timber were "merest guesses" until Henry Gannett of the Geological Survey compiled the data and estimated the area of woodlands, merchantable timber, and timber by species. However, his work lacked the specifics necessary for accurate management forecasts. More precise knowledge about smaller units of national forestland was needed.[44]

Reconnaissance—the on-site determination of tree species, reproduction, damage, soil types, and undergrowth in the national forests—began in 1908. It was the "first systematic step" toward the regulation of the federal timberlands.[45]

A typical reconnaissance crew included a chief of party, estimators, compassmen, a draftsman, and, if these jobs were not rotated through the crew, a cook and a packer. For countless miles and endless months, through all types of weather and terrain, they worked out of temporary camps, scattered across the Wild West. Their object was to examine, map, and estimate timber stands in the national forests. A reconnaissance might be either "complete" or "cursory." The latter, also known as the "ocular method," initially involved the close study of sample circular plots within 40-acre blocks of timber. A. B. Recknagel told a 1908 meeting of the Society of American Foresters that the latter type was the "standard all over the country" for the Forest Service.[46]

David T. Mason explained the procedure in 1910 as it was being followed in District 1. Estimates of each 40-acre tract were recorded on a topographic map "showing all stands of young growth as well as the areas of merchantable timber." Along with those data went a record of age and density, tree volume according to diameter at breast height, and the number of merchantable logs. The estimators—or "cruisers" as they were called—determined the percentage of mature and defective trees on each 40-acre plot that could be cut in ten years and in twenty-five years, and the percentage to be cut "in case a sale involving the area is made."[47]

Reconnaissance involved more than estimating timber. When Aldo Leopold led a crew on the Apache National Forest in Arizona, he complained about the problems involved in his job. After discovering that a previous surveying error of one thousand feet had made some of his data inaccurate, Leopold asked for an additional seven days to make corrections. He was already five days behind schedule, but he thought his many other duties justified the additional delay. "My work," he explained, "has already been interrupted by two days of necessary Timber Sale work and three days by the Sheep Springs Fire."[48]

By 1914, District 1 had standardized its procedures in a *Manual for Timber Reconnaissance.* It included, in addition to the usual instructions for mapping and classifying the timber stands, further details about reconnaissance. Each cruiser, for instance, was to carry with him an aneroid barometer, levels, a compass, a tally register, a diameter tape, and a 1-meter "cruising stick" with a useful measure printed along each side. Provision lists, tents, office supplies, and map cases were also described in the reconnaissance manual.[49]

The first step "on the long road towards the goal of having working plans for each Forest" was possible once the reconnaissance data began to arrive in the District offices. The working plans, or timber-management plans, would provide the basis for national-forest management. As explained by Henry Graves, once the volume, species, soil conditions, and reproduction had been determined, cutting plans were to be devised "so as to secure the continuous production of wood and timber." Without such plans, the sustained-yield goal could not be determined with reasonable certainty.[50]

The Division of Forestry acquired considerable experience in planning while making studies under Circular 21. Crews assigned to the studies, which included most personnel in the Division, traveled to various large industrial ownerships around the country, and hundreds of 10- to 200-acre privately owned woodlots.[51] It was an invaluable exercise. Young foresters acquired firsthand experience in woods work, valuable contact with timber owners and lumber crews, and practical knowledge. More important, the Circular 21 studies produced a previously nonexistent body of data on many major species of timber. The efforts produced the first comprehensive body of information on growth and yield, reproduction, and American cutting practices.

With that knowledge in hand and with the accumulation of data from reconnaissance work between 1908 and 1910, the foresters began to work on plans for the management of national-forest timber. Estimating the extent, density, and character of young growth that is necessary to ensure future timber resources, forest supervisors sought to develop "the information necessary to properly establish the work and policy in connection with grazing, settlement, permanent improvements, and planting," although strict adherence to long-term objectives was less important than workable alternatives.[52]

One key to success would be the size of the area, or "working circle," to be included in a plan. If a circle were too large, the plan would negate the concept of community welfare; if too small, the plan would preclude the entry of larger, more efficient timber mills. Burt Kirkland thought that "regulation by watershed" was a workable method. Others objected to Kirkland's "pet theory" on the grounds that his division was too small and that it would be more logical to draw working circles around established or potential lumber markets. George Cecil, Kirkland's district forester, agreed with the watershed idea, however, and he defended it in the interest of community welfare.[53]

Ovid M. Butler of District 4, on the other hand, favored the market theory of timber management. In areas of Utah and southern Idaho, local consumption made up only an insignificant percentage of the market for national-forest timber. Plans based upon that demand could not place the forests under complete sustained-yield management. Butler proposed first to "meet present and future demands of the strictly local markets and then to dispose of the surplus annual cut in the general competitive market." He believed that "markets largely determine the most practicable and profitable scheme of management." They were prescribed by transportation routes. Salt Lake City, Boise, and Idaho Falls should therefore be considered "market blocks," around which District 4 management plans evolved. That concept of management by markets—or meeting local needs first and then selling the surplus to the general market—would subsequently be adopted throughout the Service.[54]

One requisite for timber-management planning was reliable data on growth and yield. According to George Cecil, "growth studies show us what severity of cutting is advisable, what approximate diameter limit should be used, and how many trees per acre must be left in order to secure the maximum growth." Without accurate projections of growth and yield, a timber-management plan was no more useful than no plan at all.[55]

Gifford Pinchot provided a ''simple'' method of determining yield in his *Primer of Forestry:* count the trees on an acre, cut down an average tree, and find the tree's age by counting the growth rings. Then accurately measure the tree, divide the volume by the age, and multiply the product by the number of trees in order to obtain yield per acre. Of course, he warned, ''it is unfortunate that the simple and easy process is not always reliable, because it is hard to find either an average acre or an average tree.''[56]

Contributions to growth and yield studies were made by H. H. Chapman and J. G. Stetson in 1909, but it was not until Forest Service timber sales approached significant levels that a more reliable standard was achieved. Researcher Edward J. Hanzlik advanced the importance of site quality as a factor in predictions of growth and yield. Since the productive capacity of land is a function of soil fertility and climate, surmised Hanzlik, the ability of the land to reproduce should be predictable. Hanzlik tested his theory by studying stocked forestlands; he found three distinct site classifications.[57]

The accuracy of volume tables was another subject that required detailed study for planning use. As Henry Graves explained in 1903, there were over forty different volume tables in use across the United States. ''Many of them are defective,'' he said, ''and some are almost absurd.'' The same forest plot, when calculated according to some of the more popular tables then in use, could yield astonishing variations in the amount of timber. Such wide variations, critic A. L. Daniels maintained, caused a ''great injustice to the sellers'' and gave them ''a certain pretext for recrimination and fraud in a vain attempt to get even with the buyers.''[58]

Graves and Pinchot were among the first American foresters to study log content systematically. One of their first projects was a volume table for spruce, based on measured log content. Later they produced a study of white pine in which they used a ''form factor,'' or the ratio between the diameter of a tree inside the bark at a certain height and the diameter outside the bark at breast height.[59]

Other contributions to the development of timber measurement evolved from the Circular 21 projects. Perhaps the one with the most persistent influence on Forest Service policy, not only for its contribution to volumetric technology but also because of its application to many forestry-related subjects, was Edward T. Allen's study of red fir (Douglas-fir) in 1903. Allen measured ''thousands of trees'' felled in logging camps and made ''careful estimates of standing timber,'' studying ''the relations which age, height, diameter, and volume bear to each other.''[60]

Ideally, Service foresters would have preferred to concern themselves less with actual log content and to sell timber on the stump as is, but a federal agency must demonstrate consistent and accurate performance. As log-rule inventor Judson F. Clark explained, circumstances required a compromise in policy so as to appease both the buying and the selling public in the same transaction. ''If the Forester had his own way,'' Clark noted wryly, ''he would discard the product in all log measurement, substituting a volume unit with a classification of the logs measured into three or four diameter classes.'' Then, purchasers could do whatever they wanted with the logs, including ''burn them.'' But ''for the present,'' he admitted, ''we bow to usage and content ourselves with evolution where we would gladly see revolution.'' As a result, the Forest Service developed management plans around the timber-sale policy of estimating and then selling for harvest only that timber for which it could predictably forecast the lumber yield.[61]

Checking and rechecking of volume tables was an unending task. ''To determine as accurately as possible the rate of appreciation,'' investigator Louis Margolin said in 1906, ''the Forest Service has been conducting a series of experiments known as Mill Scale Studies.'' Those began with the measurement and individual marking of each tree felled on a sale and ended with a tally of every board sawed from the identified logs. Such ''natural'' problems as sapwood, clearness of the wood, rot shake, and worm holes required that volume estimates be readjusted. Estimates of volume were seriously hampered by ''artificial'' problems—including the amount of product left on the stump, the amount of top lopped off to rot in the woods, the skill employed in cutting logs, the care in trimming branches, the efficiency of the sawyer (a ''most disturbing factor''), ''overrun'' (production above the predicted volume) at sawmills, the judgment of the lumber inspectors, and market demands. The best that Margolin could advise his readers was that ''forestry is not a mathematically accurate science, and fair averages are all that may be expected from forest measurements.''[62]

A few imprecise adjustments also found their way into policies, if and when local logging custom dictated. One such variation involved the standard measurement of logs in 16-foot lengths. In the Pacific Northwest and in parts of the area of the northern Rockies called the Inland Empire, 24-foot logs sometimes were the standard by which private timber was bought and sold. A potential for conflict existed when purchasers expected to pay for only three 24-foot logs in a tree, and the Forest Service intended to sell five 16-foot logs with the expectation of complete utilization and full payment.

Criticism of Forest Service policy was intense in the region. J. P. Hughes, a northwestern lumberman, wrote to Pinchot to voice his contempt for a 16-foot measurement, which, he said, "borders closely on absurdity." After all, he continued, log measures were nothing more than an "arbitrary set of rules" established by lumbermen for use as a "local uniform standard." His advice to the Forest Service was to adapt its system until shorter lengths came into use and until logging equipment had been adjusted accordingly. A later decision to alter some formulas to accommodate 24-foot lengths "was based not so much upon any published or written data as upon a desire to do away with criticism and conform as much as possible to local custom."[63]

Other complaints from the private sector did not achieve the same results. Many objections to the unfairness of Forest Service Timber estimates were raised by purchasers who believed they were at a disadvantage. In the interest of checking on the reliability of its decisions and of affirming the procedures used in timber-management planning, the Forest Service investigated many of the more serious charges but seldom found itself at fault.[64]

SYSTEMS OF TIMBER CUTTING

Sales and measurement procedures were not the only considerations in timber management. There was also the question of how trees should be cut. Methods imported from Europe were divided into two basic types of cutting systems—clearcut and selection.[65]

The type of cut that was selected for a given forest was important, because each species of tree requires certain conditions for reproduction. In brief, some seeds sprout only under full shade, making them "tolerant" of shade, while others seek sunlight and are therefore labeled shade "intolerant." Those that require direct sunlight, the "intolerant" species, are best harvested by cutting "clear," or "clear-cutting." Among intolerant species subject to seed wind throw, natural reproduction is attained only if large areas are completely opened for the germination of windblown seed from adjacent stands. In varieties not subject to wind throw, a few well-placed seed trees or strips of adjoining trees may achieve the same effect after timber harvest.

Selection cutting, on the other hand, is considered most applicable to tolerant species. By selecting only mature trees for harvest and by leaving the younger growth, shade conditions foster the establishment of new generations. However, species and site susceptibility to blow down, mistletoe infestations, and other dangers are also important considerations. Some shallow-rooted species might better be clearcut on moist sites where there is danger of blow down and be selected on sheltered or dry sites. On the other hand, greater sunlight on south slopes can permit some intolerants to thrive in partial shade.

Mixed in with those reforestation issues are several combinations. Some species, for example, take root under the shade of mature trees but do not reach larger growth until the overstory has been removed. An accumulation of forest debris, or duff, may be all that is required to achieve reforestation in some varieties, while others depend upon the burning off of duff, or the scarifying of the soil, in order to take root. In fact, with some species, fire is essential in order to release seed from protective cones.

Complicating the timber-management considerations are various other factors that may influence the cutting of a timber stand. Such events as a major fire may render a forest even-aged. A minor infestation of timber-destroying insects or diseases may skip through a forest and leave an "all-aged" or "uneven-aged" stand in its wake. Similarly, the presence or absence of particular species may also prescribe levels of timber management. Such conditions exist when several species succeed each other in the life of a forest. Over a period of centuries, according to forestry theory, a forest evolves to a state known as a "climax forest," in which a single species or forest type (a certain community of species) dominates, or remains, until it is removed. In between will be found several species or types of greater or lesser value than the climax. Depending on which conditions provide the most utility, foresters may manage with the object of terminating or hastening conditions that lead to a climax state. Whether a forest is even-aged, uneven-aged, or stabilized, forest conditions influence the volume, extent, and type of cutting that takes place before sustained yield is achieved.

Interest within the Division of Forestry about harvesting techniques in the United States led Henry Graves to canvass private timber owners on their methods of harvest. He received one thousand replies describing practices ranging from selection to variations on clearcutting with or without seed trees remaining. It was apparent at the turn of the century that no systematic procedures had yet been devised for harvesting American species, at least with much attention to silvicultural needs or concern for future harvests.[66]

Determining the proper cutting system was a political as well as a silvicultural question, even in 1900. The Division of Forestry recommended cutting practices in the Adirondacks after it received a request, under Circular 21, from the state of New York. The division proposed to achieve the object of "sustained and increasing income" by selectively removing old trees above a 12-inch diameter. The replacement of hardwood species by "more valuable softwoods" was also considered but was dropped by the division as not being economically practical at that time.[67]

The study did not receive much notice until the state of New York authorized $5,000 for the reforestation of previously cut lands on the Cornell Demonstration Forest. Bernhard Fernow intended to convert the broadleaf forest into an ''economic forest'' of conifers and to use selected clearcutting as the shortest path to his objective. Unfortunately, the funds were not sufficient to allow him to cut in small, unobtrusive patterns. That led him to larger cuts, which further slowed his progress, because of his inability to plant among the ''large masses of brush-wood,'' and increased the fire hazard. Fernow reported ''the complaints of influential adjoiners regarding the so-called forest devastation, the illegal use of state forests, and the damage to their hunting interests increased.'' Instead of providing an example of economic forestry, Fernow lost his teaching job at Cornell and ended up arguing his case in court. Some of the contrary testimony came from ''experts'' in the Forest Service, who argued that selection was the correct system for the Cornell forest.[68]

Justifiable professional differences of opinion existed among the foresters who debated the Cornell case. Fernow committed several technical errors: he failed to develop a working plan, he should have cut smaller parcels, and he should have experimented first. More important was his political ineptitude, however. A thoroughgoing missionary forester, Fernow chose his goal and went after it armed with his zeal, wrapped in the righteousness of his cause, and oblivious to everything else—including the opinions of the forest's neighbors. As Pinchot—himself a paragon of righteous, crusading forestry—observed years later, the project was ''inexcusably mishandled.'' It offered lessons, however, that the Forest Service failed to take to heart.[69]

E. T. Allen's study of Douglas-fir represented the first major effort in the United States to catalog the silvicultural characteristics of a species. The Bureau of Forestry selected Douglas-fir, according to Allen, ''although other American trees are at present of equal commercial importance and in greater danger of extinction, [because] few promise to exert such influence on the lumber supply of the future.''[70]

The Douglas-fir that Allen examined was characterized by old-growth timber, some with 200,000 board feet to the acre. After studying thousands of acres between Puget Sound and southern Oregon, Allen concluded that Douglas-fir was an intolerant species that required clearcutting. Because of its tendency to seed only where the ground was clear of duff, he also recommended the burning of slash and debris following logging. That, too, he noted, would destroy the competitive hemlock species, which sprouts in the rich humus of the forest floor.

Proceeding one step further, Allen studied the rate of growth of Douglas-fir and concluded that the species reached its largest annual growth at around seventy-five or eighty years of age. ''The greatest production of wood,'' he calculated, ''can be secured by cutting every ninety years.''[71]

Allen's observation that seed from Douglas-fir could be airborne for up to a mile proved to be only partly valid. By 1909 the Cascade (now Willamette) National Forest was attempting to overcome a lack of regeneration by ''leaving patches or strips of seed trees and burning the cut over area to furnish the required seed bed for red fir.'' One year later, researcher Thornton T. Munger opined that ''the selection system seems preferable to a clear-cutting system'' for the ''fog belt.''[72]

District Forester George Cecil proposed a new plan for Douglas-fir management, but he did not discuss Munger's recommendation. Based upon a rotation of one hundred years, Cecil prescribed the division of working circles into twenty-year age classes. By proceeding through the ''older and more decadent stands first,'' Cecil expected the ''proper distribution of age classes and the normal yield'' to be reached in one hundred years.[73]

William B. Greeley disagreed with Munger's proposal to adopt selective cutting in Douglas-fir. He proposed instead that ''clear cutting followed by artificial restocking would be the best and most consistent silvicultural system.'' Austin Cary concurred, telling Greeley that ''it is clear to all that [with] a large share of the timber under the care of the Forest Service, clear cutting is the proper method of treatment.'' Burt Kirkland was of a similar mind. He theorized that Douglas-fir required clearcutting because strip cutting was too expensive for railroad logging and because seed trees were subject to blow down. That may have been enough to keep early plans for the management of Douglas-fir timber on a clearcutting basis, but it is more probable that an abundance of ''decadent, over-mature'' trees, estimated at 73 percent of the total volume in District 6, was the determining factor.[74]

The management of lodgepole pine in the northern Rockies was also in an experimental phase. Original harvest plans for that intolerant species called for a seed-tree arrangement with ten to fifteen seed trees in a group. When windfall took its toll, Pinchot suggested alternating blocks of cut and leave. His seed blocks fared no better in heavy winds than did the small seed groups, however. The District's chief of silviculture then experimented with the cutting and leaving of alternate strips, which experienced some success.[75]

In 1910 the supervisor of the Beartooth National Forest proposed to clearcut some 10 million board feet of lodgepole pine. He was ''reasonably certain'' that reseeding would take place by cutting the valley bottoms and lower slopes and leaving the upper slopes for seed. His counterpart on the Deerlodge National Forest acknowledged using such clearcuts, and he reported that the system "is apparently working well."[76]

Little was known about the timber management of ponderosa pine in 1908. That species, called western yellow pine in the early literature, ranges from Canada to Mexico throughout the intermountain region. Tolerant in the first year or so and intolerant thereafter, ponderosa pine tends toward many-aged, or uneven-aged, stands. Because of that tendency, selection cutting was recommended in the early years.[77]

Regardless of the cutting systems selected for each species, the Washington office told the Districts to manage with a view toward subsequent harvests. The Service had told the public that sustained yield would result from the management of national forests. Harvests that might belie that promise were kept in check by the directive to "leave on a sale area enough timber to justify a second cut in 30 to 50 years." "This means," the Washington office emphasized, "usually [leaving] about one-third of the timber now merchantable."[78]

The difficulty in adapting European systems to American conditions became more apparent with each increase in Forest Service timber sales, modest as those increases were. As sales grew in volume, the potential for irreversible errors increased proportionately. Whether a tree was tolerant or intolerant was immaterial if the harvest precluded the eventual restocking of the land. In that context, a more scientific way of ensuring reforestation became necessary.

TIMBER RESEARCH BEGINS

Franklin B. Hough was, in the 1870s, the first to propose the establishment of forest experiment stations in various parts of the country. His suggestions presaged Fernow's pronouncement that "no more efficient means of education in the practical arts . . . can be devised than the establishment of *experiment stations.*" In 1902 President Roosevelt created two experimental forests in the Nebraska Sandhills. Under the direction of Charles E. Bessey, professor of botany at the University of Nebraska, the treeless expanses were planted in ponderosa pine. Those first "man-made forests" established the possibility of growing trees wherever natural conditions proved suitable.[79]

The Bureau of Forestry, in an attempt "to demonstrate the feasibility of conducting logging operations [so as] to leave the cut-over forest in the best possible condition for future growth," signed an agreement with the Madera Suger Pine Company in 1904. Forty acres of the company's land were set aside for experimental purposes. "The real value of the experiment,'' E. A. Sterling said, ''must lie in the attempts to produce with one cutting, without greatly added expense, conditions that are generally favorable to volunteer young growth.'' The possibility of artificially restocking the land was another ''interesting experiment'' intended by Sterling.[80]

Raphael Zon, chief of the Office of Silvics, traveled to Europe in 1908 to visit forest experiment stations. On his return he offered a plan to duplicate the European stations for ''experiment and studies leading to a full and exact knowledge of American silviculture, to the most economic utilization of the products of the forest, and to a fuller appreciation of the indirect benefits of the forest.''[81] Zon then went to Flagstaff, Arizona, to select a location for the first station. Near the San Francisco Mountains he laid out the boundaries of Fort Valley Experiment Station. ''Here,'' Zon reportedly proclaimed, ''we shall plant the tree of research.''[82]

Gustav Adolph Pearson arrived at the Fort Valley station in August 1908. ''I was put in charge,'' he wrote later, ''and here I was destined to spend practically all of my professional life, studying ponderosa pine.'' It was, in fact, for all of his professional life; Pearson died at his desk in 1949, writing his final manuscript, *Ponderosa Pine in the Southwest*. His first project was a preliminary study of ''why western yellow pine failed to restock after cutting.'' *Reproduction of Yellow Pine in the Southwest* was published as Forest Service Circular no. 174—the first of many experiment-station publications intended to give Forest Service timber management a scientific basis.[83]

Other experimental forests and research stations appeared as the Forest Service made good on its avowed intention to establish locations in each District, with substations where necessary. ''Empirical methods,'' Assistant Forester Cox explained to the Districts, ''such as have heretofore been used almost exclusively cannot be depended on to settle beyond question the problems which are met in forest management, and intensive research methods, such as are possible only at regularly established stations, must be adopted.''[84] With the opening of the Forest Products Laboratory at Madison, Wisconsin, in 1910, the Forest Service was well on its way to a major research program. The results would be slow in developing, however. The reasons for that were twofold. Research data on species with life spans of a century or two could not be hurried, and as Pearson said, the first ten years must be devoted to ''providing the necessities of life as a prerequisite to effective research.''[85]

Forest Service research eventually became wide-ranging, but it was heavily influenced by the agency's world view. The prevailing scientific wisdom of the late nineteenth and early twentieth centuries held that a forest, once destroyed, could not easily replace itself—except with human intervention. That belief underlay the fear of timber famine, and it supported the Service's enduring obsession with forest fires, which it believed to be as destructive as the worst kind of logging. The research program therefore tended to concentrate on reforestation, including artificial replanting.

Scientific wisdom evolved, however, as did nature. The denuded part of the East reforested itself during the twentieth century, belying predictions that it could not. Moreover, science, especially botany, moved from the old evolutionary traditions reflected in plant succession—epitomized by the concept of the climax forest—into the study of dynamic interrelationships represented in ecology, which addressed each landscape as a complex of many things. The Forest Service, however, clung to older beliefs, its fear of timber famine undergirded by early scientific concepts and by the idea that wood was the foundation of civilization.[86]

The reforestation of cutover and burned lands was therefore a high priority from the start of the forest reserves. General Land Office records show some experimental seeding in California, but efforts to sustain the work were plagued by lack of funds. "There are many places," the *Forest Reserve Manual* explained in 1902, "where an expenditure of money for that purpose may, at some future time, be warranted. Generally, however, the great extent of the reserve forests, the limited market, the uncertainty of success in the very places where the method is most called for, indicate a conservative attitude, and not until a far more liberal appropriation of money is assured is there reason for introducing seeding and planting methods on more than an experimental scale."[87]

The Forest Service was more optimistic. Five-year planting programs began in 1908. In the first year, they planted 884,827 trees, 5,164,673 seedlings and transplants in nurseries, 5,000,000 seeds in nurseries, and 635.5 pounds of Douglas-fir seed in the Pacific Northwest, and 61 species of eucalyptus experimentally in Southern California.[88]

The foresters held that timber-management plans were of no consequence if fire, insects, or disease ravaged a forest. The importance of fire control led to a separate division within the Forest Service, but the lack of funds for research on insects and disease forced the Service to seek help elsewhere. Until 1912, rangers battled pests on their own. At that time the Forest Service initiated an agreement with the USDA's Bureau of Entomology. "Intensive investigations," forest officers were told in 1914, were to be handled by the bureau, with forest personnel ''strictly'' confined to securing information that would make the bureau's work ''most efficient.''[89]

READDRESSING TIMBER APPRAISAL

Congressional pressure on the Forest Service to become self-supporting, along with the desire to institute intensive management, encouraged increased timber-sale activity after the reorganization of 1908. Sales receipts passed $1 million in 1910 and nearly doubled over the following decade, although in proportion to the size of the forest system and the national timber economy alike the activity was fairly modest.[90]

The rate schedules adopted by the Washington office were unpopular. Neither the timber industry nor the Districts approved of minimum, intermediate, and maximum rates. Few operators were willing to enter long-term contracts that they thought were based on arbitrary and unprofitable stumpage prices. Even after Franklin H. Smith had opened the Chicago sales office, the gap between federal foresters and the timber industry widened as retail lumber prices continued to decline. If the Forest Service wished to do business, it had to accommodate the business community. Appraisals were the most difficult and politically sensitive part of the timber-management equation. They also were one part that the Forest Service could not address without reference to people outside the agency.

District Forester Arthur Ringland established the Lumbermen's Advisory Board in the Southwest to learn more about the needs of the industry. Similarly, in California, District Forester Coert DuBois held a conference with the presidents of several lumber companies, who expressed very little confidence in Forest Service appraisal methods. ''Just what is your basis for arriving at initial stumpage prices?'' the lumbermen wanted to know. DuBois confessed that he was ''far from sure'' what it was. He understood how the Forest Service was attempting to estimate logging and milling costs and to measure them against ''a very hazy figure'' called ''lumber values in the vicinity,'' but the determination of how Graves and his staff had arrived at one of three ''accepted figures'' for stumpage escaped DuBois. District 5 engaged in ''too much undignified dickering'' with timber-sale applicants, he complained. ''The question is simply,'' DuBois said to Graves, ''are we to charge all the traffic will bear, or are we to set a price . . . that will allow a fair margin of profit to the operator?''[91]

District Forester Edward A. Sherman in Ogden was no less dissatisfied with the ''dissimilar, slip-shod, and unreliable'' attention paid to appraisals in his District. ''For instance,'' he rebuked his forest supervisors, ''in many cases, thorough and complete estimates of logging and milling costs or the determination of the selling prices of lumber were not made, no uniform rule was followed as to the profit operators should be allowed, etc., the result being inequalities in stumpage rates and profits as between operators working under similar conditions.''[92]

Staff officer James W. Girard saw the same from Missoula. ''Up to about 1911 or 1912,'' he reported later, ''the stumpage value of Government timber was either arbitrarily fixed or roughly and often inaccurately determined from a superficial examination of the area.'' A like opinion prevailed in the Pacific Northwest. Most sales there, an officer declared, were a ''disgrace,'' and ''the less said about the way they were administered the better.''[93]

The need to control competition between national forests had been a major factor in the decision to establish rate schedules within a District, but that was not effective when Districts competed with other Districts. Such was the case in District 5, where Douglas-fir imported into California by sea from Washington and Oregon cost less than local timber shipped by rail, despite the fact that the local timber was of lower quality. A similar situation arose in District 4, where sales were appraised for the Boise market at higher rates than those used in District 6. Timber from eastern Oregon, including the Whitman National Forest, was reported to have forced twenty-four Idaho mills either out of business or into idleness. ''Without a standardization of methods of appraising timber,'' Ovid M. Butler pointed out, ''stumpage rates in the two regions can not be consistent and fair.''[94]

Few individuals in the Forest Service or in the timber industry disagreed with Butler's contention that it was ''absolutely essential that the minimum stumpage rates be forgotten and that the stumpage values be worked out upon thorough estimates of the cost of logging.'' Equally few understood the many complexities associated with determining equitable and repeatable methods of assessing timber values.[95]

The idea of applying more precision to timber appraisals appears to have originated with Bernhard Fernow. In his classic *Economics of Forestry,* Fernow demonstrated the feasibility of using the formula for compound interest as the basis for determining a fair and equitable rate of return, or what he termed the ''soil expectancy value.'' That suggested a flexible method for determining stumpage values. As it happened, however, Fernow's formulas worked well enough in judging the values of land and timber for outright purchase or exchange, but they were inappropriate for sales of federal timber, purchased on a pay-as-cut basis. Nevertheless, the basic idea had some merit in accommodating business realities in a formal pricing structure.[96]

E. T. Allen's ''red fir'' study, completed in 1903, contained the seeds of what would eventually become the Forest Service's method for obtaining appraisal information. Allen studied Douglas-fir stumpage prices, labor costs, equipment expense, and selling prices in order to determine the return on investment in lumbering operations, which he described as ''so trifling'' that ''the income of the millowner is hardly more than a fair salary for his ability to conduct business.'' Allen's methods eventually set the standard by which the Forest Service would determine the factors making up a timber appraisal.[97]

A number of people in District 5 (California) also studied the relationship of operator profit to stumpage price; examples are William B. Greeley's study of the sale of Madera sugar pine in 1905 and P. T. Harris's study of the LaMoine chance in 1906. Both secured access to company books and obtained information on actual operating expenses and profits. Another noteworthy contribution to timber-appraisal theory was the appearance of C. A. Schenck's ''Forest Finance'' pamphlet in 1909. Described by H. H. Chapman as ''thoroughly American,'' it contained ''valuable suggestions for the advanced student of Forest Finance.''[98]

District 5 took the lead in profit-ratio appraisals, producing a ''Timber Sale Policy,'' apparently distributed between May 1907 and the middle of 1908. That document made a strong case for appraisals based on a residual value after deducting costs and profits from the selling prices of lumber. The report began: ''There seems to be no valid reason why the price per thousand feet of standing timber should not bear a definite relation to the price per thousand feet of lumber manufactured from it.'' District 5 concluded that ''eventually the Government should rate the value of its stumpage on the basis of what lumber will bring in the first market, or what it will cost to put it into the first market, or the difference between the two—the profit.''[99]

In February 1910, Assistant Forester William T. Cox sent a memorandum to all district foresters, with notes on suggested silvicultural policy. It was the first formalized method for determining stumpage prices in the Forest Service. The document suggested that stumpage charges should equal the value of lumber, less the combined total of the costs of manufacture, the costs of logging, and a fair profit. ''The lumber of value,'' or the selling price, it was explained further, ''is determined by outside sources of supply, such as West Coast, Texas, Arkansas, etc.'' Comment on the formula was invited, and the Washington office was not disappointed by the response.[100]

Assistant District Forester DuBois sent a memorandum to all California national forests on 3 August 1911: ''What I want to particularly emphasize is that we have no set of arbitrary stumpage rates.'' Each potential timber sale was to be considered on its own ''merits,'' and a list of items making up the cost of production was to be ascertained in each sale ''and reported as fully in the forest description, together with the average selling price of each species and recommended stumpage rate which should allow a fair profit.''[101]

A few months later, DuBois told Graves that District 5 was studying the limits of a ''fair profit.'' By subtracting operator's costs from market price, DuBois observed, the remainder represented a sum that had to be divided equitably between the buyer and the government. ''The Service must rule definitely on the percentage of this difference it will consider as stumpage,'' he told the Washington office. If Graves would do that, DuBois believed that stumpage value should be the average selling price, less operating costs, times percentage, according to a precise formula. ''If this system is adopted,'' DuBois continued, it would ''inspire confidence,'' because each operator would obtain the same percentage of profit, regardless of the chance. Of course, he added, his method did not charge for ''slip-shod methods'' or ''poor management,'' but it did allow ''the small man'' to operate on a ''competitive basis with the big mill.''[102]

''I cannot agree with the conclusions of your method,'' Henry Graves wrote in response. The Service needed a ''uniform operating profit,'' as opposed to the formula suggested. Graves believed DuBois's formula included a ''discrepancy'' that led to ''a different operating profit in practically every sale.'' He suggested a method by which ''the difference between cost of production and selling price is divided between the purchaser and the Government.'' To that end, Graves proposed the standardization of the elements that entered into operating costs. ''To make this plan specific,'' he explained, ''I suggest that an operating profit of 25 per cent on the amount invested in each thousand feet of timber, from stump to [railway] cars, including stumpage price, be allowed in all sales involving approximately average conditions as to risk.'' That meant that the average selling price was to equal operating costs, plus stumpage, plus 25 percent.[103]

The difference between the two formulas revealed an important disagreement between District 5 and the Washington office at that early date. San Francisco devised a formula so that the operator's profits varied between sales, while the Washington office calculated the same profit for each sale.

DuBois and his staff disagreed with Graves's critique of their proposal. Conceding that the District 5 formula would make it necessary to go ''deep into the financial aspect of each operation,'' they believed that ''Mr. Greeley's formula'' did not go far enough to guarantee fairness to competitive operators. The discrepancy, as they viewed it, would be apparent if an operator deposited a small sum to obtain a bid, then borrowed a large amount to complete the contract, and ''cleaned up'' a substantial profit on his real investment. That type of operation, the District believed, should receive less compensation for the investment than a company that had expended large sums of unborrowed capital.

Along with his comments, DuBois's aide T. D. Woodbury enclosed a ''keen critical analysis'' of the Forester's formula (so named because it emerged from the office of the Forester, as the head of the Forest Service was known at that time) by Forest Assistant Swift Berry. Berry agreed that the Forester's formula was a ''big step'' toward accounting for the capital invested by operators, but he suggested: ''It seems to me that there is a danger of it not covering the entire investment. . . . Stock in the yard, cash on hand for transacting business, and similar factors,'' should also be considered. In place of the Forester's formula, Berry proposed a study of ''the average yearly capital necessary to operate successfully under average conditions'' and to establish ''proper rates'' for the ''degree of risk involved.'' Believing that stumpage ''must be placed in the reach of operators,'' Berry offered a third formula, based on the average profit per thousand board feet: Stumpage equals selling price, less operating cost, less {average investment rate divided by annual cut}.[104]

Graves answered both District 5 memorandums together. Regarding Woodbury and DuBois's point distinguishing between borrowed and unborrowed capital, Graves believed that the source of capital should have no bearing on Forest Service appraisals. The ''business ability'' to use capital ''advantageously,'' he suggested, was the factor in ''industrial enterprises which always earns the larger share of the profit.'' Likewise, Graves believed Berry's formula was unsatisfactory because it was ''equivalent to fixing an arbitrary profit in dollars and cents per thousand board feet which presumably would be applicable to all purchasers. . . . To deal fairly between competing purchasers, I feel it is essential that we stick to the basis of a profit which represents a return upon the investment.'' In his opinion the so-called Forester's formula accomplished that objective. Graves concluded that the profit fixed by dividing the selling price of timber by a percentage, and then subtracting operating costs, gave a return to the operator that ''compared favorably with returns from any other form of industrial investment with which I am familiar.''[105]

While the Washington and San Francisco offices were corresponding, appraisal suggestions emerged from other Districts. One of the first was included in Ovid Butler's District 4 ''Market Plan.'' Butler was concerned with the level of difficulty involved in the investment method. Save for a maintenance charge, Butler thought it was ''utterly out of the question'' and ''absurd'' to include the operator's investment in mill and equipment in any timber appraisal. An appraiser's "sense of responsibility" was impaired, he believed, when those items had to be determined with accuracy. Butler suggested a formula wherein stumpage price would equal selling price, less operating costs, less {operating costs divided by percentage profit}.[106]

Graves agreed with Butler's observation that it took a "high degree of skill" to make an accurate estimate of investment. He also believed that it was "the duty of the Forest Service to attain it," and he emphasized, "I do not feel that we can properly appraise stumpage values until we are able to estimate with fair accuracy all of these factors which enter into the proposition." Graves therefore rejected Butler's recommendations to exclude charges for the cost of the mill, logging and railway equipment, interest, taxes, and insurance.[107]

Dorr Skeels in District 1 had two objections to the Forester's formula. In the first place, he thought it did not do what it purported to do—that is, base the operator's profit on investment. Second, "it was not possible" to combine logging and manufacturing expenses as a basis for calculating profits. "Cost of manufacturing," he explained, "should in no way form a basis for profit whereby the profit is increased as the cost of manufacture increases." On the other hand, logging operations deserved higher profits as the level of difficulty increased. With that in mind, Skeels submitted "our Profit formula" to Washington. It held that the stumpage price should equal the average lumber-selling price, less the cost of logging and milling, and less (for profit) 10 percent of milling cost, less 20 percent of logging costs.[108]

Along with Skeels's memorandum, District Forester Ferdinand Silcox included his own suggestions. One of his main objections to the Forester's formula was that it deprived the federal government of its unearned increment—the increase in value that accrues to property from such factors as rising prices, speculation, or inflation—and that, he believed, was enough to warrant the plan's "elimination from further consideration." In its place, and like Skeels, Silcox considered the manufacture of lumber separately from the logging of timber. He believed that the stumpage price should equal the average selling price, less the cost of operation, less 10 percent of {total of cost of manufacture, cost of logging, logger's profit, and stumpage}, less 20 percent of the cost of logging.[109]

William B. Greeley replied that the Forester's formula was proposed as a "starter," with the aim of furnishing "a simple, easily-applied working method, with but one variable factor, viz: the percentage profit to be allowed." Greeley said it was not Graves's purpose "to require the use of any inflexible formula throughout the Service or to attempt a formula adapted to all conditions." Having said that, he also disagreed with District 1's contention that manufacturing and logging expenses should be separated. Since the market price of lumber, not the division of profit between logger and manufacturer or the market price of logs, had been selected as the basis for appraising Forest Service stumpage, Greeley reminded District 1 of the reasons behind the decision to base appraisals on the market value of lumber, which were:

1. The lumber market was the final determining market in timber values.
2. The lumber market was less susceptible to manipulation, to arbitrary prices, or to arbitrary or even fictitious specifications.
3. Sales contracts that were adapted to the lumber market yielded closer utilization and scaling.
4. In 80 percent of the Forest Service markets, a log market did not exist.

"The Forester's formula," Greeley said, "was designed primarily to meet the requirements of sales for long contract periods to new plants, where there is no division between logging and manufacturing of any practical significance and where very large investments in mill, railway, and logging equipment solely for the handling of government timber are involved." Other than that proviso, the Washington office had "no objections" if District 1 should choose to use either the Silcox or the Skeels formula, so long as the Forester's formula was used as a "check."[110]

William B. Hunter, special examiner with the Bureau of Corporations, undertook a study of the valuation of federal timber in 1912. Believing that none of the proposed Forest Service appraisal systems afforded "a sound basis for comparing one lumbering operation with another," Hunter avowed that "the only basis on which such comparisons can be intelligently made is the percentage of profit per annum on the amount invested." Furthermore, he believed, the extraordinary acumen, luck, or ability that reaped extra profits for talented individuals could not be measured in an appraisal. Only the "average man's" abilities could be dealt with in an estimate, Hunter told the Forest Service on 5 May 1912.

Hunter began with the question, "What constant yearly amount is required to pay a constant rate of profit on capital in use and to pay off a given part of the cost of the plant during the operation?" Proceeding from that point, he constructed a table for depreciation, based on the original investment, the expected rate of amortization, and profit. Hunter's resulting formula meant that the stumpage price would equal selling price, less operating costs, less "X."[111]

District 6 had its own opinions on the appraisal question. With Burt Kirkland and W. H. Gibbons suggesting methods similar to Hunter's formula, District Forester George Cecil presented Portland's "total investment method'' to the Washington office. Like its counterpart in the intermountain region, the District in the Northwest separated logging costs from milling costs.[112]

Assistant Forester Greeley approved of District 6's approach to appraisal theory. ''I like the general form and makeup of Kirkland's formula,'' and ''the discussion of the investment profit method in Gibbons' memorandum is excellent, in many respects the best we have yet had,'' he wrote to the district forester. Still preferring the Forester's formula, Greeley authorized Portland to use its method in sales involving large investments and long cutting periods, so long as it checked the results against the ''standard, or Forester's formula,'' and tested Hunter's formula in the process. He told Cecil: ''I have come to the conclusion . . . that in the larger, more permanent operations where appraisals are made with exceptional care and by the best experts we have, the 'investment profit' method is the sounder and more consistent.''[113]

A year before, Greeley had called the Forester's attention to the relationship between lumber-market values and the objectives of the Forest Service. Stumpage prices that were based on logging operations, he believed, resulted in 10 percent ''poorer utilization'' of the timber, because logging operators would be loath to pay for and transport marginal material if they were paid only for the volume that arrived at the mill, as was the usual practice. In addition, Greeley advocated the averaging of lumber-market prices from several years past, rather than the use of prevailing market prices. Pointing out the drop in prices between 1907 and 1911 (from $14.40 to $11.12), he argued that it was ''unwise'' for the government to base stumpage prices on the lower recent values.[114]

Forester Graves did not initially approve the averaging of market prices over several years. ''The use of a two or three year average,'' he observed, ''might easily reduce the increase under normal conditions which we would otherwise be entitled to use.'' A month later, Graves rescinded his decision and notified the Districts that ''the adoption of the average of two years' lumber prices seems advisable in order to follow more closely the normal movement of lumber values and avoid both extremely high and extremely low fluctuations.''[115]

The periodic readjustment of stumpage prices was one method suggested for the recovery of the unearned increment in long-term sales of Forest Service timber. The Washington office chose to set a figure of $2 per MBF as the dividing line. If prices did not increase above that figure during the stated readjustment interval, no change would be made; if the price rose over $2 per MBF, the Forest Service would add 75 percent of the increase to stumpage. The $2 exemption was allowed because of the "very low rate of profit" included in the initial appraisal, and if the price went over $2, 25 percent of the unearned increment would be passed to operators "to cover increases in the cost of production." Graves later confirmed the 75/25 split and established the policy of readjusting stumpage every three years in areas where the industry was "well established" and every five years in "new regions" where the "encouragement to operators" and the "development of new regions hitherto unexploited" was necessary.[116]

Another important issue was the value of an operator's plant at the close of a contract. When additional timber was available to the mill owner at the end of a cut, the plant and equipment could be expected to have a fairly high "residual value." On the other hand, when no future use of the investment was expected, the operator was entitled to a relatively low "wrecking value" in the initial stumpage appraisal. Which of those values was selected could have a significant impact on the operator's profits and, hence, on the Forest Service's sales receipts.

The Forest Service accounted for residual value in its appraisal work before 1912. It thereby assumed an obligation to sell additional timber to the operator at the completion of a contract—an "obligation which had been asserted not infrequently" in the past "because of investments made in logging improvements." That was something that Graves, compelled by law and sensitive to charges of fostering monopoly on large-volume timber sales, wished to avoid in the future. Perhaps no other element of the appraisal equation presented such a "knotty problem," Greeley said.[117]

Concern for those "obligations" led Graves to suggest that any residual value should be excluded when making appraisals for new plants constructed solely for the harvest of federal timber. "I believe," he wrote the district foresters in 1912, "it is wise to avoid such obligations by charging against the first body of timber bought an investment or sinking fund sufficient to cover the entire outlay for improvements, deducting only what actual sales value they have at the end of the contract regardless of continued operation in the same location." For existing plants, which presumably had private timber at their disposal, Graves believed such operations were not entitled to a depreciation or investment charge. "The first cost of such plants can be fairly assumed to have been paid back in the private operations for which they were built," he concluded.[118]

Graves's decision brought strong objections from some of the Districts, to which Greeley responded. He acknowledged that there were problems inherent in Graves's approach to depreciation, but he also believed other concerns made it imperative for the Forest Service to take a "liberal" attitude toward charging off depreciation on investments as an operating cost. The first concern, he said, was the need to encourage sales: "It is urgent that sale revenues be increased during the next two or three years at least to a point sufficient to pay the cost of the Forest Service." Second, the Service must relieve itself from "obligations" to the purchaser as "public policy," even though it was not "good business" to do so. In that regard, he added that our "large contracts" had already brought charges of "promoting monopoly" by "favoring certain large operators." Greeley concluded that "the entire investment should be charged off, with the exception of wrecking values." He further advised the Districts to "reserve" adjacent blocks for other purchasers, so as to avoid charges of favoritism in large sales.[119]

A final issue during the discussion over appraisal formulas was the lesser, or "inferior," species. DuBois brought the matter to Graves's attention. After pointing out the loss that operators took on species priced higher than the difference between the selling price and the operating cost, DuBois complained that the practice "works against the silvical good of the Forest—which, after all, is the Government's first interest." His point was well taken. Purchasers typically absorbed a loss on the less valuable timber species in a multispecies sale.[120]

Graves explained the reasons for the "arbitrary minimum price" for each species. It would be "difficult to explain" to other purchasers and the public why a species sold higher in one area than another. It was his opinion that "so-called inferior species" had an "intrinsic value" to the public by virtue of their future value. Until those species could carry their own cost, the Washington office would continue to set minimum sale prices for all species, even though the figures were offset by reductions in the rates of better timber and, therefore, were "partly fictitious." In conclusion, Graves said, "The more valuable species have got to carry the less valuable ones, and I do not see how any method of calculating stumpage rates will eliminate this fact."[121]

The Districts often objected to minimum rates on inferior species. Some believed it would be better to reserve those species from cutting until market value could pay the operating costs, while others believed the inclusion of inferior species would thwart timber sales. The policy stood nevertheless.[122]

AN APPRAISAL MANUAL AT LAST

Before an appraisal formula became Forest Service policy, two major theories were offered as the best means of pricing federal timber. The first, known as the Forester's formula or the operating-profit plan, an "overturn" method, approached the problem from the viewpoint of administrative practicality. This method was considered convenient for training personnel in timber-appraisal procedures.[123]

The Districts opposed the Forester's formula on two grounds. They believed it would result in having too much of the unearned increment go to the purchaser. Conversely, they found it difficult to attract buyers for large tracts, such as those requiring the construction of railroads, except when it was possible to demonstrate a return commensurate with the investment required. Only the "investment profit" formulas, among the several proposed, accomplished that. Aggravating the Districts' relations with the Washington office over those issues were occasional inter-District disputes, as when District 4 and District 6 disagreed on the pricing of timber going to the Boise market. Clearly, a uniform policy must be promulgated for the entire Service.[124]

In 1912, H. H. Chapman suggested factors that he considered important to federal methods of stumpage appraisal: "It should be the purpose of the Service first to determine the profit per thousand [MBF] available for payment of interest on the investment and then to ascertain if possible, whether this 'gross' profit is sufficient to pay a rate of interest that will compensate the investor." Chapman believed that only through an investigation of the latter factor would the Forest Service discover "the balance wheel that will govern the demand for government stumpage."[125]

Chapman's message was circulated widely, but his was not the only opinion of its kind. Austin Cary also had definite views on the investment-profit formula: "I think over what I have learned in the fir region, that if I were personally responsible for an appraisal I should feel that the thing to use, because the surest and most definite thing I know of, is the profit in money per MBF." Cary also stressed the importance of remembering the purpose behind appraisal. He cautioned that the final determination of what should or should not be included should rest on whatever "meets our purchasers on their own ground, and enables us to trade easily and clearly with them."[126]

Until the Washington office imposed an unarguable standard on the Districts, they must solve the appraisal problems on their own. District 5 was the first to publish a formal handbook on stumpage appraisal, in 1913. Swift Berry, its author, recommended the Forester's formula for the national forests in California. In large sales, however, where "good sized" investments were involved, Hunter's formula was to be used as a "counter check." Forest personnel were advised that "as Forest Service appraisers develop skill in determining the items of investments, it is felt that this or some other investment formula will prove the logical way of determining stumpage." District appraisers were to calculate the "average cost of production," based on "normal efficiency" of labor, equipment, and supervision. In small sales, where mills had ''inefficient labor and methods,'' the appraisal was to be based upon ''the methods of logging and manufacture actually applied by such mills.'' Then, each appraisal officer was advised to follow a series of detailed steps to arrive at a final figure.[127]

District 4 also issued its appraisal directives in 1913. Stumpage prices were to be based on the Forester's formula, and operators' profits were to be calculated at 20 percent when no risk was involved, 25 percent when accessibility and logging and milling costs were a ''considerable risk,'' and 30 percent when those factors were ''very great.'' Conversion costs, or ''all legitimate costs involved in converting the standing timber into lumber or other manufactured products,'' were specified by category.[128]

The activity in the Districts caused the Washington office to step in with a uniform approach. Greeley sent proposed appraisal instructions to the Districts early in 1914. Most noteworthy were the decisions to eliminate interest as an operating cost, to exclude the stumpage price from operating costs in the Forester's formula, and to adopt the investment method, using three modified methods for small sales, as the standard for appraisals by the Forest Service. Along with a request that each District comment on the proposals, Greeley was ''particularly desirous'' of obtaining ''the frank criticisms of some experienced operators'' on the instructions—the first hint in some time that the Forest Service's leaders might want to know what the affected public thought before it imposed a final procedure.[129]

The Districts made little effort to canvass timber operators, however. Instead, the Forest Service's first general instructions for timber appraisals on the national forests were the product of internal discussion. That seventy-page document, prepared by Greeley and Cary, appeared on 12 November 1914. The manual governed the appraisal of stumpage in cases of timber sales, settlement, trespass, free use, and land exchange. Each appraisal officer was allowed variations, but changes were permitted only if the standard instructions were used as a double-check on the results. The instructions, said the Washington office, were ''for the exclusive use of Forest officers and will not be furnished to persons outside of the Forest Service''—a telling reflection of the agency's outlook.[130]

Three appraisal formulas were adopted by the Forest Service from the many suggested during the preceding three years. The ''investment method,'' which basically came from the Berry and Hunter proposals, was to be the Forest Service standard. It was defined as a ''percentage return on capital invested.'' The instructions said that the new standard formula ''should be employed uniformly in appraising the larger chances, and in appraising the smaller bodies of timber where it is applicable.'' It did not, however, apply to smaller chances, where the investment was minor and the operator was the only, or the most significant, part of the labor force. Called the ''overturn method,'' the small-chance formula was basically the earlier Forester's formula, which allowed a profit on the percentage of overturn, ''or the entire production cost of a thousand feet of timber at the date of sale, including operating charges and depreciation, but not stumpage.'' A modification of the overturn method was approved for those Districts in which, it was believed, separate calculations were necessary for logging and milling operations. Each phase of production would then be offered a profit ''adjusted to its peculiar conditions and risks.''[131]

The new appraisal instructions were based on the ''prudent operator'' concept. A profit of 15 to 20 percent was considered an adequate return on investments when the life of a plant could be expected to last through the contract. As the level of investment increased in proportion to the estimated difficulty of logging according to the nature of the sale area, the authorized profit level increased to 18 to 20 percent, or even to 22 to 25 percent when it involved new facilities in unimproved areas. Each item that was considered to be a legitimate operating cost was also explained in the instructions. Appraisal officers learned that ''the chief purpose of this detailed enumeration is to cause appraisers to study all features of an operation thoroughly and take all necessary costs into account.''

The Washington office had also taken a different view of wrecking and residual value by that time. The handbook said that ''if no additional timber can be handled by the plant, it is obvious that a wrecking value only will remain.'' On the other hand, ''where additional bodies of stumpage are available, a residual value should be credited to the fixed investments.'' The Forest Service thereby avoided the ''obligation'' to reserve timber while also offering initial investors the best possible chance for winning subsequent bids.[132]

It had been three years since Coert DuBois had complained about ''too much undignified dickering'' in timber sales by the Forest Service. Each of the Districts and many pioneer foresters took part in the discussions leading to the resolution of the many issues and questions. It was a highly complex subject that demanded answers in the face of contradictory pressures to reach self-supporting status, to achieve the highest level of silvicultural management, to promote community stability, and to avoid charges of fostering monopoly. Nevertheless, as far as the federal foresters were concerned, they had done the job. The last element of the timber-management program was in place. Only time would tell if it would work in the real world.

THE FOREST SERVICE AND THE WORLD

The preparation of the appraisal manual reflected much that was positive in the Forest Service's collective character. Arriving at an appraisal system was an exceedingly difficult task, a practical matter far removed from "forestry as propaganda." Others who were less committed to the conservation concepts that the forestry profession had inherited might well have taken a more convenient approach to the problem.

On the other hand, the Forest Service had done the job almost wholly on its own. Although gestures were made toward consulting the timber-buying public, the Service actually followed its own muse. Its Progressivist faith in its own expertise inspired a technocratic belief that no one else could offer anything valid on affairs that were properly and wholly the Service's own. That was a product of the larger forces that created American forestry, of which the Forest Service was the overwhelmingly dominant part. Forest conservation had arisen in the beliefs that wood, more than anything else, was the foundation of civilization and that the nation would need more and more wood as it grew and prospered. Thus the supposed approach of timber famine promised calamity.

The Service held firm in its resolve to provide more wood for the nation, despite the nation's declining per capita need for wood. The Service started its own research program to study the mysteries of the forest, but from the outset that program fell behind other sciences. It perfected a timber-sales program but did not produce the massive increase in sales that it believed to be essential. It proselytized forest conservation, but the nation had grown tired of hearing about it.

The world around the Forest Service was changing, and the agency resisted facing it. As much as the federal foresters wished to embrace beliefs and principles they thought were eternal, things were not so simple. Challenging times lay ahead.

CHAPTER THREE

Selling Timber in an Uncertain Market

It is surprising to see how, even under the pressure of their great war, the French retain their bargaining instincts and their thrifty way of always providing for the future. . . . [They] regard us as wasteful in our use of wood and doubtless think that if they hold us down hard we can get on with much less than we are asking for. Also, they are taking no chances on exhausting their forests and being put to it for an adequate supply of wood after the war.

—William B. Greeley (1917)

The Forest Service needed reinvigoration, although its people would be the last to admit it. Still shaken by Pinchot's dismissal in 1910, on the eve of World War I the organization was only slowly healing under Graves's leadership. It seemed an agency with conflicting purposes—trying to become self-supporting by means of timber and grazing receipts, it had failed to do so partly because of an insufficient demand for national-forest timber and partly because its policy was aimed mostly at the future, meanwhile selling only to meet local shortages and otherwise avoiding direct competition with private industry.[1]

The agency had prepared for future demands, however. It had gathered a lot of information about the national woodlands and had developed workable procedures for the appraisal and sale of timber. The research program acquired separate status in 1915, promising future returns for resource management but separating research from administration with effects that then were unpredictable.[2]

The demands of the future lay in the future, however, and were accordingly and distressingly unknowable. For the moment, the Forest Service was shaken by repeated controversies, sniped at from all sides, sometimes appearing friendless. To broaden public support, the Service began to investigate the development of recreation just before the war. Even that became a battle for political survival, one that was partially lost when the National Park Service Act of 1916 introduced a competing conservation agency and an implied doubt of the Forest Service's ability to manage lands for recreation.[3]

There followed a decade and a half of subtle change in Forest Service programs, but scarcely any adjustment in its outlook. The agency marketed its timber aggressively, and sales increased enough that they became the means by which careers were advanced. The larger market remained doubtful, however, and by the start of the Great Depression, it collapsed altogether. The Forest Service finally was forced to reduce its sales to keep from making things worse for the timber economy. In the interim, however, the Service made some adjustments in its relations with industry, and it momentarily set aside its crusade for regulation. But the old fire—born of the fear of timber famine—smoldered beneath the surface, scarcely touched by a challenging new interpretation of "sustained yield."

CONFLICT ABROAD AND AT HOME

The dispirited organization needed a cause in order to revive the crusading spirit of its first years. World War I provided just such a spark. The experience of the war years was a watershed for the Forest Service, as it was for the forests, forestry, and the forest-products industries. It worked changes that had significant effects on the management of national-forest timber. The Forest Service's program was not only revitalized; it was also subtly redefined, reflecting altered attitudes and new leadership.

The war was a turning point for American forests and for the industries and professions dependent upon them. Mobilization converted the timber industry from a migratory consumer of virgin forests to a network of stable regional enterprises engaged in long-term production. The experience showed that even in an emergency, forests did not face inevitable and irreversible destruction: second-growth timber established itself in the market, and forests and forest industries bid fair to have a long future together. The war also reduced old tensions between timber industries and professional foresters, thus abating arguments for regulation and promoting cooperation.[4]

The first effects that the European war had on American forest industries in 1914 were negative, as foreign purchase orders were canceled. Within a few months, however, orders for forest products increased astronomically, because the Allies lacked the manpower to exploit forests that were not in the hands of the enemy. Production, employment, and profits in American industry increased dramatically, enabling it to support the campaign for "Preparedness" and, ultimately, the United States' entry into the war. The demands were unprecedented. Millions of board feet of lumber were required for training camps and other facilities, railroad cars, war materiel, crates and containers, and a thousand wooden cargo ships. Demand for national-forest timber, however, was the last to develop. Except for overgrazing, the national forests experienced little direct impact from the war. The time when their timber would be in high demand was apparently accelerated, however.

The timber industry fell under increasing federal control during the war, as did nearly every other sector of the economy. The experience strengthened industrial trade associations and increased the industry-wide cooperation necessary to allow lumbermen to deal with the government on more equal terms. The industry also emerged from the war with an improved public image. No longer were timbermen decried as heartless plunderers of public resources; they were as cooperative and patriotic as any other citizens. And the central management of the national economy led to familiarity between industry and government. As a result, the federal foresters moderated their demands for federal control of the timber industry and turned increasingly toward cooperation.

The most dramatic event was the direct participation of foresters and woodsmen in the conflict. When the United States entered the war in 1917, the Allies projected a steady build-up for a war of attrition that would last for years. The effort demanded huge quantities of timber. Nearly the entire trench system of the Western Front was constructed of wood, shielded by earth. Millions of feet of wood were required for entanglement stakes, buildings, road-and-bridge construction, containers, docks and warehouses, and other uses, as well as for firewood.

With their manpower in the trenches, the Allies could not meet their own timber requirements. Limited shipping space made American exports impossible. Therefore, forestry units were established in the United States and Canadian armies, for service in France. More than thirty thousand American foresters, woodsmen, and mill men served in Europe. Back home, another large group of logging units produced aircraft material in the Spruce Production Division of the Signal Corps' Aviation Section. The demand devastated the Forest Service's payroll. The cream of the professional staff volunteered for military service, leaving behind a skeleton organization. A dispirited Henry Graves proclaimed the Service "a crippled organization . . . with troubles piling up of every description."[5]

He spoke too soon. Veterans of military forestry returned with a strong sense of shared experience and contribution to the war effort, which subtly influenced relations between the industry and the government. The governmental foresters who went to France, among them Graves and his successor, Greeley, had gained an appreciation of industry's perspective. They had felt frustration from controls by French foresters. Greeley recalled:

> We had many arguments with the French foresters over cutting requirements and I found myself on the other side of the table from similar controversies with loggers back home. The Frenchmen were understanding and realistic—and mighty good woodsmen. But in issues between their established regime of timber culture and exigencies of Allied manpower or speed in getting wood to the front, the forest always won out. . . . A grizzled *conservateur* said with a fatherly smile, to a bunch of impatient Americans: "Our forests have fought several wars before this one."[6]

The Forest Service softened its attitude toward industry after the war, increasingly favoring public-private cooperation. The wartime experience was decidedly a contributing element, but the change was probably occurring anyway. Graves, rebuilding his bureau after the shattering climax of the Pinchot years, had accepted the myth of the timber famine and favored the public control of private harvesting practices. But he was not about to take on bloody battles that he was bound to lose.

In 1919 he proposed a comprehensive national forestry plan. The danger of fire loomed larger than private harvesting practices in his worries about the future of the forests, however, so he called for an expanded state-federal cooperative program of fire protection. Not altogether incidentally, his legislation would require the states to regulate logging in a manner approved by the Forest Service. Graves hoped to buy the favor of industry for his form of regulation with an expanded fire-protection program. Industrial spokesmen dismissed the plan, however. Worse, Gifford Pinchot formed a committee of the Society of American Foresters to review the state of the nation's forest resources. Not surprisingly, Pinchot's group predicted the "beginning of timber shortage." Just as predictably, Pinchot's committee demanded federal regulation of all forest lands. To Graves, Pinchot's plan contained "many socialistic features." Said Pinchot, "The continued misuse of forest lands privately owned has now brought about a critical situation in America."[7]

Timber management on the national forests threatened to become enmired in yet another political wrangle. Pinchot would fail to have his way, however, and not only because of his abrasive and extremist rhetoric. The Society of American Foresters' endorsement of Pinchot's proposal sparked a series of debates in the organization, revealing that no longer did foresters speak with one mind on political questions.

More fundamentally, the nation voted firmly against reforms and crusades and in favor of "Normalcy" in 1920. Regulators like Pinchot were advancing notions that were unacceptable to the nation and were bound to be declared unconstitutional by the Supreme Court of the period. The national interest in private-forest regulation was nebulous.

William B. Greeley, third chief of the Forest Service. A pragmatic man who attempted only the possible, he sought cooperation between the Service and the timber industry, from both sides of the issue (courtesy Forest History Society).

Child labor was, in contrast, a universally condemned evil. However, in 1918 the Supreme Court denied Congress the power to eliminate child labor used in interstate commerce, and in 1922—in a case arising out of the forest-products industry—the Court forbade the regulation of child labor by means of taxation. In such an atmosphere, legislation to regulate private forest lands would have been a waste of time.[8]

Graves retired early in 1920; he was succeeded by William B. Greeley. Above all, Greeley was a pragmatic man, inclined to attempt only the possible. Nor was he persuaded that the timber industry was an unbridled menace or that unregulated harvesting posed the greatest danger to the nation's forests. His philosophy earned him the bitter enmity of Pinchot, but Greeley proved able to cooperate with industry. His ways met with decided approval among those in the Forest Service and in forestry who shared his practical bent.[9] A decided changing of the generations had taken place in Forest Service leadership.

One of the things that worried Greeley was the danger to the Forest Service that constant political dogfighting posed. The survival of the agency, let alone its plans for timber management, repeatedly appeared to be at stake. A case in point was Alaska, upon which the Forest Service placed high hopes. A vast area, rich in resources and short of population, Alaska promised an opportunity for a positive demonstration of the value of national forests. Moreover, it was the one region in which the Forest Service's receipts regularly exceeded its expenditures.

Greeley and Associate Chief E. A. Sherman hatched a plan to make Alaska a "second Norway industrially," by the coordinated use of natural resources. In 1921 the Forest Service issued a prospectus for 335 million cubic feet of pulpwood on the Tongass National Forest, to be logged over a period of fifty years. One sale of a billion board feet had a contract over a thirty-two-year period, with five-year reappraisals; according to a press release, it was "in line with the general policy of the Forest Service for making available the timber resources of Alaska as a means of increasing the supply of pulpwood for the United States." The purchasers were to build a 100-ton pulp mill, and future sales to keep it running were promised. Several other large sales followed over the next two years.[10]

The Alaska program ran into considerable controversy, because its scheme was grand but was poorly explained to the public. Assistant District Forester Fred E. Ames laid the groundwork with a study of pulpwood industries in British Columbia; he concluded that the United States was bound to repeat the same mistakes in Alaska. Before 1921, both areas offered timber on extremely liberal terms but without safeguards for local economies. High transportation costs and failure to look after long-term stability had wrecked a number of industries in the Canadian province. The Forest Service's new program for Alaska was intended to avoid those deficiencies, but repeated rejections of proposals for industrial development brought the agency castigation from parties interested in exploiting Alaska's timber.[11]

Before the new Alaska program could get under way, Secretary of the Interior Albert B. Fall launched an attempt to transfer the national forests to his own department. He began with the Alaskan forests, and when he failed to obtain transfer by executive order, he turned to Congress. Counseled by Secretary of Agriculture Henry C. Wallace, Greeley resisted with a Fabian strategy, willing to wait out the storm. Pinchot, however, interpreted Greeley's attitude as concurrence, so he challenged Fall in public. The result was a lingering dispute that kept the Forest Service's status in doubt, until President Warren Harding publicly stated his approval of the Forest Service in the summer of 1923. The issue finally dropped out of sight, although it may have turned less on the relative merits of the two agencies than on the political disgrace of Secretary Fall.[12]

COMING TO TERMS WITH INDUSTRY

Greeley was determined to put an end to political head knocking and to come to terms with the timber industry on some program that would satisfy all but the extremists. To respond to Pinchot's proposed plan of federal regulation, Greeley assembled the National Forestry Program Committee, composed of representatives from industry trade associations and from a number of conservation and fire-prevention organizations. He then prepared legislation emphasizing a national fire-protection program, a thorough piece of cooperation that would encourage better harvesting on private lands, but not at the point of the federal sword.

The industry was satisfied, and Greeley sent his plan to the state foresters and to his district foresters for support. Pinchot, most notably, stood out in opposition. He favored a fire-protection program, but he would not budge on his demand for harvest regulations. The pragmatic Greeley sought only the possible, including $1 million to start work in fiscal 1922; he did not want to jeopardize it by asking for the impossible. "It is my position," he said, "that the problem of forest devastation is three-quarters a problem of preventing fire . . . and that other measures of reforestation such as the regulation of cutting may well wait until we have gotten the forest fire measure under reasonable control."[13]

In 1920, in response to a resolution sponsored by Senator Arthur Capper of Kansas, the Forest Service produced the "Capper Report," assessing the state of the nation's forests. The report asserted that there had been a serious depletion of national timber resources, causing record high prices. Focusing on the most solvable problem, the report called for increased cooperative fire protection.

What ensued was a period of push-and-pull, including public debates between Pinchot and Greeley, which focused on alternative legislative packages advanced by Pinchot, on the one hand, and by the National Forestry Program Committee, on the other. The latter proved the more effective in aligning support and in sending witnesses to Congress; it made Pinchot appear to be a raving extremist. When Pinchot turned his energies to running for the governorship of Pennsylvania, the arts of compromise produced the Clarke-McNary Act, of 7 June 1924.

This act emphasized federal/state/private cooperation for the protection of forests, especially from fire, in a considerable expansion of the Weeks Act of 1911. The law also authorized cooperative federal/state nurseries, related reforestation programs, and technical advice and services to owners of small woodlots, with the states and the federal government as equal partners. Additionally, the Clarke-McNary Act expanded the kinds of land that could be purchased for national-forest purposes, and it allowed the Forest Service to accept gifts of land and to enter into exchanges with other federal agencies.[14]

Most important, the Clarke-McNary Act ended the considerable drain of Forest Service energy into efforts to control all the nation's timber. The Service now had many programs, inside as well as outside the national forests. When it came to basic activities such as timber management, it now should devote undivided attention to the lands under its own control. That happened none too soon, because following World War I, demand for national-forest timber began to increase.

THE TIMBER SALES OF THE 1920S

Norman E. Johnson, a veteran of the Forest Service in the Southwest, recalled that during the early 1920s the "rule of thumb" was to mark a million board feet of timber for every mile of railroad laid by a timber contractor; the contracts typically guaranteed two-thirds of the volume of timber in the sale area. Some timber markers believed that 90 percent of the volume was necessary if a contractor were to break even. Consequently, as timber sales increased during the 1920s, there was a great deal of heavy cutting, moderated later by improvement cutting—or culling of trees to improve the quality of the future forest.[15]

The heavy timber work load early in the 1920s caused the Forest Service to consider the costs of making timber sales. In District 6, Logging Engineer James W. Girard produced, in 1920, "detailed time studies" showing the gross output of timber appraisers and markers per work hour for trees of different sizes; he supported them with a series of graphs. To his dismay, "these studies showed that the effective time per day was less in every case than the time that the men were expected to be on the job. In other words, the men were not on duty full time." He believed that timber-sales people spent too much time walking to and from the job; therefore they did not mark as much timber for sale as he believed they should.[16]

The cost of making sales was high enough that the control of expenses was vital, in government and in industry. Within a year, District 1 adopted Girard's "curves" and urged them upon industry as a fair basis for payment of timber cruisers. From that grew a system of recommended pay scales for every aspect of timber harvesting. By early 1921 Girard had extended his work to include a handbook showing the costs of teams and teamsters in harvest work, including veterinary services, shoes and nails, harness repairs, and forty pounds of oats and sixty pounds of hay per team per day. Increasingly elaborate cost studies on both the governmental and the private sides of the timber-harvest equation aimed at improving appraisals by defining every cost element.[17]

Appraisals still caused technical problems. Those in California drew criticism from the Washington office for "using different rates for working capital and fixed investment. For the sake of uniformity," said the District Office, "we should hereafter use one rate." The District was told to use "the investment method for logging operations and the overturn method for manufacturing and shipping operations."[18]

As sales went forward, Forest Service researchers tried to develop information on which to base long-term management plans. "The extensive logged-off areas of the two states [Washington and Oregon] offer great possibilities to grow a second crop of timber if only fire is kept out and Nature given a chance to reseed the ground," the Portland office announced in 1920, reflecting the earliest in a long series of regeneration and second-growth studies of Douglas-fir.[19] Meanwhile, Robert H. Weidman investigated selection versus clearcutting (he favored the latter, though few others did) in ponderosa pine, and Duncan Dunning tried to determine tree growth from the size of the crown or from the total surface of leaves.[20]

The growing program of timber sales did not stop the Forest Service from preaching to the public. In 1922 Greeley and others collaborated on an attempt to explain to the public their concept of the place of forests in American society. Called "Timber: Mine or Crop?" their contribution to the annual *Yearbook of Agriculture* helped to seal the agricultural analogy in Forest Service policy—that is, forest resources were "crops" that could be cultivated and harvested perpetually, as were other agricultural commodities. They contrasted that philosophy with "timber mining," or the cut-out-and-get-out practices of the past. Timber mining, they said, reduced the native stand of United States timber from 5,200 billion board feet to 1,600 billion board feet of virgin timber, to which were added 600 billion board feet of culled and second-growth timber—half the total in the three Pacific Coast states. An original forest extent of 822 million acres had declined to 138 million acres of virgin stands and 250 million acres of comparatively inferior culled and second-growth timber, not to mention 81 million unproductive acres.

The foresters also talked about the future. The lumber industry was gradually shifting to the Far West, raising prices in eastern markets and leaving behind cutover lands that would require more than thirty years "before we can make our forests produce through growth as much timber as is now yearly taken from them, and a period of shortage is inescapable.'' In other words, the Forest Service would do its best, but it would not accept blame for the timber famine it had predicted for so long. Their prescription for what ailed America's forests was, first, ''effective protection against fire'' and, second, reforestation, at least by leaving seed trees in harvest areas. Mostly they advised a stop to uncontrolled exploitation, the reduction of waste, and an increase in timber production to full capacity. They said the national forests would do their part, and they predicted that under intensive management, those forests could produce 6 to 8 billion board-feet of lumber. ''The alternative is idle forest lands and timber bankruptcy.''[21]

Philosophy was translated into action. Timber-management plans and timber sales were couched in long terms and were accompanied by press releases extolling perpetual value and community stability. The Crater National Forest management plan prescribed the Medford Working Circle, all the forest area in the Rogue River watershed, an estimated 8,779 MMBF in fourteen major species. It was divided into seven districts according to watersheds, each subdivided into blocks, with a coherent program of long-range support for local industries.[22]

The Fruit Growers Supply Company, founded in 1908 to produce stock for fruit boxes, by 1922 was searching for dependable supplies of timber. That year it entered into a contract with the Forest Service for a billion board feet of timber on 42,722 acres of the Lassen National Forest, to be harvested in successive thirty-five-year stages. It obtained such an agreement mainly because its cutting practices on its own land ''compare[d] favorably'' with Forest Service practices and because the firm intended to remain a permanent fixture in the community.[23]

Several press releases in 1922 announced the opening of the ''largest compact body'' of yellow pine in governmental ownership, on the Malheur National Forest. ''This is in line with the Federal policy of putting the forests of our country to their highest use, instead of locking up valuable timber resources so that they are of no benefit to the American people,'' according to the Forest Service. From an estimated 7 billion board feet on 550,000 acres, the foresters promised to ''provide a continuous perpetual supply of raw material for a lumber manufacturing industry . . . capable of using from 50 to 60 million feet of logs annually.'' The first sale was for 890 MMBF on a twenty-year contract. That was followed with appeals to ''progressive southern lumber men'' to come to the area and with an announcement that ''the Government proposes to open up the region on the basis of a perpetual supply of forest products.''[24]

The billion-feet sale on the Lassen National Forest in 1922 helped the Forest Service that year to ''shatter all previous records'' for timber sales: receipts totaled $2,307,000. Lest it be thought too subservient to big industry, the Service pointed out that "not only lumber companies" secured national-forest timber. Six thousand farmers, fishermen in Alaska, coal and copper miners, railroads, pulp and paper mills, turpentine distillers, furniture makers, and a host of other small users had benefited from the public resource.[25]

As the volume of timber sales increased, the Forest Service grew sensitive to perceptions that it was in collusion with big interests, or that its methods were no different from the big-time cut-out-and-get-out lumbering it had long decried. So the agency went out of its way to emphasize the long term and its orientation toward local communities. "It is the plan of the Forest Service to offer this timber at such a rate and under such conditions that a mill may be maintained there for all time," according to foresters on the Willamette National Forest. That would be done by clearcutting and brush clearing, leaving seed trees. On the Payette National Forest, a large sale would be harvested by selection cutting in recurrent cycles, intended to attract a lumber mill "to give such a mill a permanent life." On the Plumas National Forest, the Forest Service promised "carefully" to supervise harvesting "so that all young and thrifty trees will be left for future growth." And on the Inyo and Mono national forests, "the timber from this region will be used to meet the needs of towns, ranches, and mines located in the Owens River Valley."[26]

Things were not as rosy as the press releases claimed, however. Some officials in the Forest Service began to question how well timber sales were being handled, from the standpoint of future forest production. In California, District Forester Paul G. Redington believed that too much young growth was being destroyed by logging with teams or steam engines. He proposed an investigation of caterpillar tractors, which he had heard were being used successfully in some industrial operations in Montana. Nor was marking being handled properly. Several district foresters believed that markers culled the forests, marking all the good trees and leaving nothing but second-rate material for the future. Ogden's R. H. Rutledge put it thus: "We are not developing as fast as we should the feeling on the part of the men who actually do the marking that they are responsible for the future growth of timber on the area and that this thought must dominate their actions."[27]

The difficulty in setting out and enforcing realistic standards for marking reflected the Service's primitive knowledge about the growth and yield of many species of trees. The technical literature was full of revisions, and the uncertainties in technical information translated into frequent disputes with timber buyers, who disagreed with the Forest Service's volume assessments. Marking that was too generous may have been intended to avoid disagreements with purchasers.[28]

More ominous were incidents in which the Forest Service evidenced an unseemly concern for the businesses that were using national-forest timber. That was a natural outgrowth of the propaganda campaign that "sold" the forests to the public as community assets. Supervisors of remote forests grew anxious to match the impressive sales figures of the more accessible or more valuable forests. "Loggers are getting around to our timber at last," crowed foresters at the Umpqua National Forest, announcing large cuts and even larger sales in 1923. The Forest Service had gone so far to attract buyers as to make available an entire town site on federal property, "another decided step forward in timber sale policy."[29]

Some supervisors, anxious to attract timber buyers, went too far. In 1923 the management plan for the Coconino National Forest in Arizona was revised to encourage greater annual and future production, by the device of increasing figures for the projected growth and yield of ponderosa pine by the exclusion of grazing. The forest was divided among several identified local lumber businesses, with a share being marked out for each. The objects of the plan were:

1. To grow and supply the greatest practicable amount of forest products to the general market
2. To stabilize and maintain the lumber industry on the Forest and the industries dependent on it
3. To conduct all operations so as to minimize the damage to watershed values of the Forest
4. To supply the local markets (see no. 2)

In other words, the entire national forest had no other purpose, except for a nod to watershed protection, than to look after the well-being of certain local businesses—which were in fact doing very well without using much national-forest timber. The Coconino enticed them to expand by offering public timber. Acting Forester E. A. Sherman pounced on that as soon as it hit Washington, saying: "We are interested in these [firms] as producing units rather than as business organizations, and a more impersonal attitude toward them as business organizations would be more desirable as a matter of expression." He advised revising the plan without allowing for an expansion of the number of manufacturing plants in the region, permitting only an assumption that existing mill companies would be successful bidders on any national-forest timber. While he believed that sales should be offered with an eye to the convenience of local firms, awards should still go to the highest bidder. "We should, therefore," he concluded, "be very careful about giving any assurances to specific business organizations in regard to future sales."[30]

Annoyances continued to develop as the timber-sales program advanced. Western yellow (ponderosa) pine had established itself as an important species, and high demand for it raised concerns about future production. Harvest and restocking methods were a variety of local improvisations overlain with a futile attempt to find universal prescriptions. The species was too dependent upon local circumstances, however. Researcher Julius F. Kummel demonstrated that trying to reproduce the species by seeding was a waste of time. Another investigator, W. W. White, examined selective cutting in Montana and concluded that the growth of diameter doubled and that of volume tripled in trees left after selective cutting. More in touch with the peculiarities of the species, however, was the use of three silvicultural systems on one national forest in Montana, according to the conditions of each stand.[31]

Still the search went on for a single approach to the species. In 1925, after studies were made of an area of the Wallowa National Forest cut in 1915 and 1916, Hanzlik thought he had the answer: "It may be safely followed that reserve trees which have always grown in the open and those left in natural clumps, as a rule, will show none or very little increased growth on the cutover areas, while those released or partly released by having trees cut within a radius of 50 feet or 60 feet, may be expected to show considerable increased diameter growth." But a series of memorandums subsequent to his announcement demonstrated that the differences he observed could be explained by varying rainfall and other immediate climatic conditions.[32]

After the passage of the Clarke-McNary Act, it seemed likely that the acquisition of national-forest land in the South, begun modestly under the Weeks Act of 1911, would accelerate. The Forest Service's interest in southern forests expanded, since most of the lands that it would acquire would first have been "mined" by commercial operators. Forest Service agent W. W. Ashe counseled conservation for southern harvesters, who he thought were too devoted to clearcutting in southern pine. He advised selective cutting, saying "it is wasteful of wood to cut the small trees." If smaller trees were left, he claimed, they would increase in value "at not less than 15 per cent a year for smaller sizes and at not less than 10 per cent a year for large size poles."[33]

The real action still centered in the West, however. To counter arguments that selective cutting increased the costs of logging, District Forester Redington in California sent a package of studies to eight timber companies, trying to show that it did not. Urging conservation on private as well as public lands, he said that it was "essential to the practice of forestry" to leave an average of four "thrifty young trees of seed bearing size" to restock each acre. At the same time, District 5

Railroad logging: a trainload of Douglas-fir logs arrives at the mill at Chehalis, Washington, about 1921 (courtesy Forest History Society).

considered revising its sales contracts to require that Forest Service logging engineers lay out logging railroads or trails. That would not only minimize damage to the forest; it would be more economical as well. Shortly thereafter, the same District claimed that the growing use of caterpillar tractors in logging was economically advantageous, but as practiced, it usually smashed up the forest and ruined reproduction.[34]

CHANGE IS IN THE AIR

The American economic boom during the ''Roaring Twenties'' required great quantities of forest products. The nation consumed 38.8 billion board feet of lumber in 1926, most of it for construction, but there were innumerable other uses in automobiles, furniture, fixtures, and things ranging from toys to beehives—although the per capita consumption continued to decline.[35]

Most of the nation's lumber still came from private holdings, but demands on national-forest timber were growing. Some sales had exceeded the Forest Service's ability to supervise properly, and the impetus for sales often outran that for silviculture. On one national forest a long-standing timber buyer that was generally regarded as equal to the Forest Service in its concern for the future resource nonetheless destroyed over half of the unmarked timber, two-thirds of the poles, and three-quarters of the seedlings in its harvest areas. Such statistics could be generated on most national forests.[36]

Forest Service efforts to manage its timber produced frustration. In 1927 the Oregon Legislature issued a memorial complaining that federal-timber sales "had demoralized the timber market in that state." Secretary of Agriculture William M. Jardine replied that the state misunderstood that national forest timber was sold "at not less than fair, carefully appraised prices." Explaining that the Forest Service's objectives were to stabilize existing local industries and the communities that depended upon them, to prevent the depreciation in value of federal timber, to aid the development of regions or communities by the establishment of payrolls and transportation facilities, and to harvest ripe or deteriorating timber before its value was lost, the Secretary declared: "In all cases, the starting of a new crop of timber by proper provisions while harvesting the old is an essential feature of the department's policy." He went on to note that the national forests contributed only about 5 percent of Oregon's lumber production and to point out that local residents "earnestly desired" every sale that had been made. He saw no reason to change his department's forest policies.[37]

Not all was negative, of course. The discipline of forestry had become so popular that by 1927, twenty-three institutions offered a degree in forestry and about fifty others offered courses in forestry subjects. Equally heartening were evidences that the timber industry had been infected. The National Lumber Manufacturers Association announced in June 1927 that the Camp Manufacturing Company of Franklin, Virginia, had decided to continue its operations with a view to "continuous production." Said the press release, "Consulting foresters have laid out the general plan of operations, which is mainly one of selective cutting, in place of the system of clean cutting heretofore used."[38]

The Forest Service itself was becoming more sophisticated and was thinking about matters other than timber. The management plan for the Coconino National Forest was revised in 1927, for a number of reasons. Previous studies either had neglected to account for part of the forest's timber or had been "over-conservative" in calculating growth on cutover stands. On the other hand, regeneration had been less than expected on thousands of acres that had been denuded before the forest was reserved, and generally the silvicultural prospects were not so simple as previously believed. The forest needed some more-sophisticated programs, but the situation was complicated by its earlier encouragement of local development, characterized by occasional dramatically large sales, as opposed to a steady flow of timber to local industry. The forest supervisor now wanted to concentrate on improvement cuttings—the selective harvest of diseased or insect-infested trees and the like—but he must not do so in a manner that would ruin the local businesses. "Owing to the heavy investment in mills and railroad and to other economic considerations," he said, "it is essential to maintain a steady rate of cut rather than a large cut for a time and then reduction."

In other words, the earlier parceling of territories among companies had placed the Forest Service in an awkward position if it wished to change its silvicultural management of the forest's timber sales. Even now, two of the companies were about to invade each other's territories. The solution was quietly to drop the territorial system and to let the marketplace work its way, even if it were to bankrupt one of the firms. "It will be the policy not to interfere with the free play of competition in awarding timber offered in the territory of the two operations or to favor one operation against the other, as to which shall continue in business."

The Coconino case demonstrated the dangers of placing too much emphasis on economic benefits to local industries so as to justify the existence of the national forests. Once an industry became dependent upon a national forest, its vested interest could be seriously impaired (and people could be thrown out of work) if the forest's management program changed to correct previous errors or to account for newly recognized resource values, especially those that were not so easily measured in terms of jobs and tax receipts. Nonetheless, the Coconino determined to overhaul its management plan, limiting damage to the local economy as much as possible but recognizing that an over-developed industry might have to contract. As an example of how far the new plan was willing to go to protect future silvical values at the expense of current economics, logging methods were restricted to the use of animals and steam skidders pulling logs directly to predetermined tracks. Tractors, in "an experimental stage," would be allowed when they could be used economically without damage.

Most remarkable in the new plan was the occasional subordination of timber to other resource values. "In order to protect the scenic values along roads the marking within one to five chains on either side of the road will be confined to a light selection, removing not to exceed 50% of the volume of insect infested, diseased, or overmature trees." Although that might appear to be only a grudging and incomplete accommodation of esthetic values, certain small zones obtained complete protection: "On limited areas of high value for recreation, timber production will be subordinated to recreation on which no cutting will be done."[39]

National-forest timber management had come a considerable distance since World War I. It had begun mostly as plans based chiefly on social and economic theories, the pinnacle of which was the impending "timber famine." The plans had required adjustment when tested in real life. Nationally, the Forest Service had decided that it could not reform the world by regulating private lands. Favoring cooperation, it had decided that it should set a good example. Locally, neither species of trees nor local communities had proven to be uniform around the country or responsive to simplistic solutions. The enthusiasm that had almost huckstered sales of federal timber during the early 1920s had by 1928 matured into a deliberate attempt to adjust to local realities and, when possible, to avoid irreversible decisions. There was also a glimmer of understanding that the future of neither the world nor of the forests depended on timber alone, but on other things as well.

The Forest Service was small and dispersed enough that it could exhibit a great deal of localism and a fair degree of flexibility. Foresters practiced a young art, and they knew that their discipline remained in the developing stages. The engines of change that had started during World War I were still at work, albeit less dramatically.

More change was in the air. On 30 April 1928 the McNary-Woodruff Act authorized $8 million for land purchases under the Weeks Act. The next day, satisfied that he had achieved his major goals, William B. Greeley left the Forest Service for a career in private industry. And on 22 May the McSweeney-McNary Act granted Forest Service research its own charter, separate from the National Forest System, and authorized a forest survey.[40]

THE TIMBER BUST

The lumber market peaked in 1926, then began a downward spiral that would end only with mobilization for World War II. The industry was in serious trouble by 1928. It was apparent that there was a substantial overproduction of timber, which in small but significant part included that coming from national forests under sales arranged in the preceding years. With demand falling, production eventually would fall as well.[41]

The drain on the nation's forest resources, in other words, was receding. But Gifford Pinchot picked that very time to start another quarrelsome campaign for federal regulation of private-timber operations. After leaving the Pennsylvania governor's chair early in 1927, he threw his rhetorical weight behind legislation that would have taxed all timber harvested on private land $5 per thousand board feet, with a possible rebate of $4.95 if the timber were cut in accordance with federal standards—something that most probably would have failed its first test in court. The next year he entered an informal alliance with George P. Ahern, a former Army officer and a founder of Philippines forestry, and helped him to produce a tract called ''Deforested America.'' It charged the entire timber industry with the destruction of America's forests and charged Greeley and the Forest Service with complicity in the alleged outrage.

Pinchot had alienated too many conservationists with his extreme positions and his abuse of those who disagreed with him. At the 1929 meeting of the Society of American Foresters, he could assemble only a few stalwarts to issue a ''minority'' report that dissented from the sentiments of the society's Committee on Forest Policy. The report claimed that the rate of destruction of forests had increased by more than half during the previous decade, and it again demanded central control of forest industries.

Conservationists were appalled, one of them calling Pinchot and Ahern's plan a ''socialistic and destructive document.'' Others called it a specimen of the ''timber famine bogy.'' The National Lumber Manufacturers Association predictably opposed Pinchot's notions (answering ''Deforested America'' with a pamphlet called ''Reforested America''). When the august American Forestry Association disagreed with the self-appointed defender of the trees, Pinchot condemned the association as being a lackey of the industry, claiming that its magazine sought to ''facilitate the continued destruction of American forests.''[42]

Pinchot's renewed escapades had little effect on the Forest Service's political programs or on its dealings with the industry, with which it had developed a cautious working relationship. The declining demand for timber caused problems far more immediate than any timber famine, but it also meant that the Service might be able to increase its attention to other aspects of silviculture as sales programs fell off.

When District 4 proposed an ambitious ''selling campaign'' to prop up its declining volume of timber sales, E. A. Sherman of the Washington office advised caution. Suggesting that sales be limited to small size only for ''local consumption,'' Sherman and others wished to avoid ambitious large programs without knowing what effect they would have on present timber producers in the District. The District's grand plans would depress the industry further unless sales were limited to ''present operators.'' When the Ogden office asked what it could do to raise receipts and cut expenses, Sherman said that it was more important to undertake timber-stand improvement and to train forest officers in silviculture.[43]

Despite the old dream of controlling all the nation's forests and despite the long-standing salute to community stability, the Forest Service had given relatively little attention to forests that surround

David T. Mason, consulting forester. He challenged the Forest Service with a new interpretation of sustained yield (courtesy Forest History Society).

national forests during the heady times of big sales and growing receipts. Facing a depressed timber industry and falling sales volumes, federal foresters looked increasingly to the communities of which they were a part. To a great extent, the Forest Service came under the influence of consulting forester David T. Mason and his partner Carl M. Stevens, the preeminent advocates of sustained yield.

A NEW LOOK AT SUSTAINED YIELD

Sustained yield had been at the heart of the entire federal forestry campaign from the beginning. It was rather a vague notion for many years, simplistically asserting that the nation would not run out of trees if one were planted to replace each one that was logged. Mason shifted the emphasis from the forests to the forest industries, making sustained yield mean the continuous production of forest products, rather than forests. Conservation should conserve the industry and therefore its supply of raw material—particularly that within a reasonable distance. National supplies meant little to local producers, who depended upon local supplies. Mason intended to sustain local supplies by drawing on public and private sources. In other words, national forests should be integrated into the entire timber-producing region that supported given local economies.[44]

Mason and Stevens were persuasive and sophisticated, and because they advised careful harvesting, the Forest Service was receptive. Stevens told District Forester Rutledge in Ogden early in 1928 that his objective was to use forest property for the "greatest possible net gain." All forests "must be selectively cut," he said, to get "the material of greatest value." He envisioned three systems of selection—by area (according to accessibility), by tree species (according to market value), and by size (according to grade and handling qualities). He claimed that "tremendous profits" were possible in such a system. As Mason and Stevens encouraged loggers to eschew the total consumption of forests, they also asked the Forest Service to be more flexible about what trees should or should not be cut.[45]

Mason told Rutledge that he and William B. Greeley were in "complete agreement" on Mason's theory of sustained yield. He said that Greeley was "willing to use the national forest timber . . . to promote the general adoption of sustained yield by private owners but he is willing to do this only where public opinion would now permit." He also claimed that Greeley was willing to authorize sales of fifty to sixty years duration "where necessary," even if cutting could not begin for a decade or two. Mason said further that by "great good luck" he had been able to present his ideas to Secretary of Commerce Herbert C. Hoover, who, he suggested, would help to "create the public opinion" necessary for the adoption of his plan. Hoover thought that it would be the solution to the current overproduction of timber.[46]

Rutledge was an easy convert. He told Stevens that timber operators who were considering sustained-yield plans in southern Idaho "need not necessarily think in terms of privately owned land alone, but there can be worked out such a thing as sustained yield or continuous operation on lands owned partly by the operator and partly by the Federal Government or State." Stevens's "selective cutting principles will fit in admirably in this conception."[47]

The idea spread quickly and took hold. National forests were to be managed as part of the management of a wider area in order to sustain the local forest industries; national-forest timber was but part of the timber that required consideration if the stability of the industries were to be ensured. Witness the new *Forest Management Handbook* for the national forests of California:

> Sustained yield cutting is the goal in every case and this principle will be adhered to except when such action would result in unwarrantable silvicultural sacrifices during the first cutting cycle. In those working circles where the government has control of the situation the sustained yield principle will be complied with even though such action has the result of deferring cutting for many years to come or until economic conditions permit of profitable operation at the sustained yield rate.
>
> Within working circles where the probable purchases of the National Forest stumpage controls, or probably will control, as much as 50% of the total stumpage, sustained yield may often be brought about by working out and presenting to the private owner a cooperative sustained yield program providing for the leaving of the necessary minimum amount of timber on private lands and fair reimbursement for real values left on the area thru land exchange.[48]

The forest industries of western Washington were deeply depressed by 1929. They accordingly attracted the first practical experiment in sustained yield, as Mason and others in the private sector envisioned it. The Western Forestry and Conservation Association published a cooperative forest study of the Grays Harbor area, which was quickly followed by another survey paid for by the businessmen of Elma. The result was an assessment of the forest resources of Grays Harbor and Mason counties and a plan of systematic exploitation to maintain economic stability. Greeley, who by then was secretary of the West Coast Lumbermen's Association, thought the plan demonstrated that public timber was essential to ''bridge the gap'' of twenty years before private lands would again be productive. Christopher M. Granger, speaking for the Forest Service, was skeptical. He suggested that there was not enough stability in private management to achieve sustained productivity, so local industries must be dependent upon national-forest timber alone, which would be ''sold strictly on a sustained-yield basis.''

The planners were persistent, and they had thought of everything. For example: ''A vital factor that should receive earnest consideration is the closer use of material now wasted; present logging operations . . . leave approximately twenty cords of wood of pulp size or better to the acre.'' Caution and persistence eventually paid off in the establishment of the Shelton Cooperative Sustained Yield Unit, under Forest Service auspices, after World War II.[49]

Industrial logging practices in the great pine forests of the West were indeed such that the Forest Service had a right to be skeptical about the private sector's ability to change its ways. Even on the national forests, standard practices were brutal. In many areas of ponderosa pine and Douglas-fir, the standard harvest technique before

1930 was "railroad logging"—essentially the building of a railroad through a valley and the cutting out on either side as far as yarding cables would reach. It required 40,000 to 60,000 board feet per acre to pay. Contracts usually called for reserving a third of the merchantable timber, but the residual varied depending upon whether it was believed that a particular volume of seed trees was required for regeneration.[50]

The Forest Service permitted railroad logging because it was economical, because there was reason to believe that the forests would recover, and perhaps because it had always been done that way in some regions. The sizes of clearcut areas were astonishing by later standards. Forest Service Chief Edward P. Cliff recalled having seen railroad logging in the Northwest around 1931: loggers "clearcut all the timber that could be reached from the logging railroad with the equipment that was then available," in a pattern that dated from the start of the national forests.[51]

If the Forest Service allowed big harvests to go forward in such a manner under its control, its dim view of unregulated practices on private lands could be questionable for certain species. A certain "if we do it, it's right; if they do it, it's wrong" attitude infected some research studies. Examinations of public and private (selectively) cutover plots in southern Oregon produced the usual varied findings. According to the author, "The habit of yellow pine to grow in many-aged stands makes growth computations a very complicated problem." Variations in growth occurred on both public and private lands, but any growth on the latter he attributed to "economic conditions rather than silvicultural practice."[52]

Other studies of ponderosa pine during the late 1920s proved equally inconclusive in the way of generalizations about growth and yield. They focused on selective harvesting, but what *selective* meant in any given circumstance was an uncertain commodity. It was difficult to urge the Forest Service's selective-cutting methods on private operators when the Service's methods seemed variable, giving results that were equally variable.[53]

The large-scale, clean-sweep logging of ponderosa pine and Douglas-fir, on private and public lands alike, contributed to the overproduction that was undermining the timber industry in the Northwest. Equally apparent was the fact that most of the easy timber had been harvested; what remained was more difficult to get out because of forbidding terrain. Both the Forest Service and the industry examined the economics of logging according to techniques and equipment. They aimed to preserve it as an economic enterprise and to reconcile harvesting with the Service's selection for regeneration. To logging engineer Axel Brandstrom, the modern tractor was the answer: "It is seen that caterpillar tractors which are instruments for intensive selection 'par excellence,' occupy the coveted position of showing in nearly all cases the lowest cost for logs of all sizes and for all yarding distances." William B. Greeley thought the salvation of the industry lay in a return to methods of forty years earlier, before railroad logging, again to enhance "intensive selection."[54]

In the final analysis, by 1930 the Forest Service's methods were undefined and utterly up in the air, at least for the great western pines and firs. According to the agency's own Walter H. Meyer, the Service was committed to "a selection method of cutting in order to perpetuate the supply of timber and to ensure a future stand. Whether or not the method of cutting is actually a true selection system is a question which currently goes the forestry rounds."[55]

Such questions were about to become academic. As the depression deepened, the Forest Service continued to push timber sales, to the dismay of a particularly depressed industry. Sales could not even meet Forest Service purposes, however. A typical case occurred on the Coconino National Forest in 1930, when a sawmill quit cutting timber but held its contracts open for ten years or more, effectively tying the hands both of the Forest Service and of anyone else who might have been able to operate successfully.[56]

Even H. H. Chapman was discouraged: "The stability and prosperity of the wood-using industries is the life blood of forestry. . . . Continuity is the only hope of retaining and building up a trained body of foresters, without whom forest production cannot become truly efficient." Industry finally acted. Wilson Compton, of the National Lumber Manufacturers Association, and other industry representatives persuaded President Herbert Hoover to order federal agencies to "go slow" on timber sales and to decline "to have Government timber used to put more operations in business in an industry already overcrowded with unused capacity."[57]

As if to explain the necessity for the "go slow" policy, Chapman reminded foresters of the new meaning of sustained yield in 1931: "In forestry, therefore, a sustained yield requires, first, that as far as possible the standing timber or forest capital be cut in such quantities that no complete break in yield will occur which is serious enough to disrupt the dependent industries and change the habits of consumers of wood."[58]

The Forest Service's response to "go slow" was to initiate the forest survey that had been authorized by the McSweeney-McNary Act of 1928. The agency had been so busy with timber sales that it had not revised the inventory of the national forests since 1922. Work was under way in the Northwest by the spring of 1930, focusing on the Douglas-fir region of western Oregon and Washington. Supervisors were to inventory timber by type and volume on their own forests. Examiners were assigned by county to compile information from ''existing records in the possession of land owners or their agents, timber brokers, and from various public records.'' Confidentiality was to be ensured.[59]

The start of the survey was both an opportunity to gather information on the state of the nation's forests and a chance to justify Forest Service programs. In May 1930 a high-level committee gathered in Madison, Wisconsin, to determine the course of the work. The Subcommittee on Requirements said the survey should determine the ''current need'' for forest products, according to a ''National Drain Table''; trends and underlying causes affecting forest products; major opportunities ''for expansion for products of our forest lands''; and an estimate of forest needs for ''noncommodity uses.'' The Subcommittee on Growth and Drain believed that growth data were ''urgently needed'' in order to calculate the drain on the timber supply. It also recommended separating data by ownership class, specifically to tarnish the gleam of the National Park Service. ''Because . . . it is important that the State, region, or Nation know how much—in timber growth—parks are costing them, it is desirable to indicate the growth rate in parks even though the wood itself may not be available for conversion.''[60]

James W. Girard transferred to the research staff in 1930 to supervise the field work for the entire country; he remained in that job until the start of World War II. His task was daunting. He must determine forest acreage and timber volume; forest depletion from cutting, fire, and insects; forest drain that was being replaced by forest growth; timber requirements for all purposes; and an analysis and comprehensive report for each national forest.

Girard perceived that the volume tables would not be finished ''until the second coming'' if he were to follow the traditional methods. He had observed, however, that there was a strong relation between what he called the ''form class'' and total tree volume. Using 32-foot logs, for instance, he found that 50 percent of the volume of a three-log tree was in the butt log, the top diameter of which was specifically defined by the form class. Following that principle, he developed a method of gauging tree volume, which he applied to 22 million acres in the South by taking quarter-acre samples at ten-mile intervals. When skeptics in the Washington office sent their own crews to double-check him with extremely slow methods, his figures proved to be better than anything short of tape-measuring virtually all trees. His method was double-checked for each region before it extended across the country.[61]

By September 1930, with the survey under way, the Forest Service felt called upon to explain what it was doing. ''The Forest Survey is not merely to find out whether there is going to be enough wood to go around,'' it announced. ''It is not simply to prove whether there will or will not be a timber famine. It is rather a major economic study with numberless ramifications.'' Its findings would affect the livelihoods of ten million people and would ''touch the national, State and private pocketbooks, bear directly on the welfare and stability of thousands of large and small communities, and call insistently for an answer before we can know how to proceed economically and effectively in meeting our forest needs and using our forest lands profitably.'' In a nutshell, and both contradicting its opening promise and revealing that the Forest Service's fundamental eschatology was unaffected by experience, the statement concluded: ''Light will be thrown on the question of whether or not there is danger of a national timber famine.''[62]

Such puffery notwithstanding, the forest survey would provide mountains of hard information upon which to base future plans and programs. That would help to end the improvisation that had characterized the timber-sale boom of the 1920s. It also reflected the fact that despite the growth of recreation and the increasing importance of watershed and grazing management, the production of timber came first in the minds of most people in the Forest Service.

NEW PLANS AND OLD ASSUMPTIONS

The authors of a new management plan for the Lincoln National Forest in New Mexico in 1931 cast a jaundiced eye at the increase of recreational visitors and summer-home builders. It also expressed unhappiness at the fact that local industries were drawing timber from Indian, state, and private lands. Ignoring the depression in the industry, the plan proposed to step up sales, with the forest's first priority being ''a permanent timber supply for local settlers.'' ''Sustained yield'' was little more than a shibboleth in the plan, which was clearly aimed at attracting customers that were already being served adequately by other landowners. Watershed protection was the forest's fourth priority (this in the dry Southwest). The Crook National Forest in Arizona did manage to make watershed its first priority, but its timber plan so disregarded the Coronado Trail (recreation was the third priority) that the Washington office called it to account.[63]

The forest supervisors wanted to recapture the glamorous timber-sale figures of the previous decade. According to Axel Brandstrom, ''doubt and despair'' gripped the private timber owners, competitors of the Forest Service. He urged them to adopt ''economic selection'' as the way out. They were not interested in any federal Trojan horses, but they did accept several Forest Service studies in regard to the efficiencies of their logging and mill operations.[64]

Meanwhile, the spread of white-pine blister rust in the Northwest threatened to upset all calculations regarding sustained yield and the maintenance of a stable industry. Ovid Butler of the American Forestry Association raised the alarm in 1932, predicting that increasing harvests of diseased trees would flood the market and would further ravage a shaky timber industry.[65]

The Forest Service, like the timber industry, was adrift by 1932. Facing budget cuts and layoffs of personnel, serving an uncertain market in a national economy that piled disaster upon disaster, many in the organization looked vainly to a resurgence of the postwar boom in timber sales as a way of focusing their energy. That was a cruel self-deception; things larger than the timber market would have to change first. As in 1917, it would take a national crusade to bring the Service out of its slump.

The prophecy of the crusade came in the form of *A National Plan for American Forestry,* the so-called Copeland Report, prepared in answer to a resolution sponsored by Senator Royal Copeland of New York, who in the spring of 1932 had wondered aloud whether a great reforestation program might relieve unemployment. The Forest Service made preparation of the plan a Servicewide effort, but without significant participation by outsiders. Contributions were solicited from all Regions (formerly Districts), which were then distilled into the final document by a large staff detailed to Washington. Proposals and language that would antagonize segments of the public were sweetened, and as a result the product received mixed reviews within the Forest Service. Elsewhere, it was widely regarded as a grand vision.

The Copeland Report was the first review of the national forest situation since the Capper Report of a decade earlier. Regarding timber, as might have been expected, it averred that nearly all the conservation problems were on private lands. It went farther, asserting that the majority of private lands were not susceptible to conservation, which could be achieved, however, with expanded cooperative programs on the remainder. Half of the nation's timber resources, according to the report, should pass into federal ownership through massive expansions of the National Forest System.

The Copeland Report was not merely another wave of the timber famine argument; it was the first expression of the concept of multiple use in its modern interpretations. It proposed significant public and private reforestation programs, and it wished for the extension of cooperation to insect and disease control and to other land-management concerns besides fires. It addressed water resources, range management, recreation, wildlife, research, aid to the states, and many other matters. And it came at providentially the right time, just as the very receptive Franklin D. Roosevelt entered the White House, in March 1933.[66]

CHAPTER FOUR

Timber Management Takes Control

Thus in the mid-thirties with timber production at a low ebb the national forests were with some notable exceptions essentially in an undeveloped state from a timber production standpoint. Because of the depletion of privately owned stands of timber to the point where only a relatively few operators owned a long term supply, the demand for national forest timber began to increase tremendously during the war years. The Forest Service was forced to accept in part the highly undesirable practice of over-cutting on some of the developed areas in order to increase the over-all supply of timber.

—B. H. Payne (1948)

Sales of national-forest timber did not dominate the Forest Service's time in the 1930s as they had after World War I, because lumber glutted the market during the Great Depression. There were other things to think about, and because of New Deal programs, the Forest Service's most glamorous activity became the control and prevention of forest fires. Nevertheless, the agency was galvanized to action by the experience of national hard times. Its responsibilities multiplied in a blizzard of New Deal programs; an unsuccessful one reawakened the campaign for regulation. That became the worst fight yet over the issue, and it accompanied vicious struggles with the Department of the Interior, wherein "multiple use" failed its first test. Quelling uncertainties in its ranks, the Forest Service next faced the test of World War II. That conflict and its aftermath made sales of federal timber an important fraction of the economy at last, and by the late 1940s the agency felt ready to meet all challenges, politically and otherwise. But if the timber program reinvigorated the Service, it also took control of its destiny.

NEW DEAL FORESTRY

The Forest Service drafted a bill to implement the Copeland Report program, and it cautiously urged its field men to mobilize political support during President Roosevelt's "First Hundred Days." The legislation, which would have extended the Clarke-McNary Act to cover forest insects, diseases, soil erosion, and floods, applied to wild and managed lands alike. Appropriations authorizations were to be increased from $2.5 million to $6.5 million, the majority for fire prevention and control. Additional provisions would have authorized the distribution of planting stock to all forest owners, expanded extension services in regard to farm woodlots, and extended the National Forest System (the latter by spending half a billion dollars, at $30 million a year).[1]

It appeared at first that the new president might diminish rather than expand the Forest Service. On 10 June 1933 he issued an executive order that consolidated the National Park System, removing from the War Department and the Department of Agriculture national monuments, military parks, and other historic and scientific properties formerly in their jurisdictions and placing them in the hands of the National Park Service of the Department of the Interior. All national monuments in national forests were suddenly to be managed by others.[2]

The Forest Service had resisted that transfer, and it took the event personally. It did not have time to mourn its loss, however, as the president handed the agency a staggering load of new responsibilities. Before long, all things seemed possible in forest conservation, for there was an abundance of funding and, most important, manpower. The tentative implementation of the Copeland Report seemed too modest by comparison.

On 5 April the president established the Office of Emergency Conservation Work. He was already moving rapidly to put the unemployed to work and to restore America's natural resources in the process. When he had accepted the Democratic nomination in 1932, he had alluded to a million-man conservation army. Immediately after taking office, he sent Congress a bill to establish the Civilian Conservation Corps (CCC). Responsibility for raising, equipping, housing, and managing the personnel of the CCC went to the United States Army, which Roosevelt told to have a quarter-million men in the woods by summer. That the army did. By the fall of 1933 the quartermasters had built wooden housing for 250,000 men—placing one of the largest single orders in the history of the lumber industry; this was easily filled from the available glut.

Responsibility for the work of the CCC, however, rested with civilian agencies. More than half of the fourteen hundred CCC camps were on Forest Service lands, and the agency put the men in the camps to work with a will. The biggest single employment of the CCC on the national forests was in fire control—not just fire fighting, but also the construction of roads, trails, fuel breaks, communications, detection systems, and so on. There was other work as well, much of it in timber management. A great deal of brush control and timber-stand thinning, formerly thought impractical, progressed during the 1930s. Considerable CCC effort went into the development of campgrounds, picnic areas, and other recreational amenities. By the end of 1933, twelve thousand CCC enrollees were working in twenty-two states on the control of blister rust, with a budget of over $2 million. The CCC is perhaps best remembered for planting trees, however. Half the trees ever planted in the United States were placed by CCC enrollees. It was an old dream come to life. Since the beginning of the conservation movement, foresters, impatient with natural regeneration, had promoted planting as the way to reforestation. It had not been a practical question until CCC manpower made it possible on a gigantic scale. That was especially important in ravaged forests of the East and South, where throughout the 1930s, Forest Service acquisitions of land for national forest purposes went forward at a remarkable pace.[3]

Other work on national forests was supported by funds from various of the New Deal's "alphabet soup" of emergency employment and conservation programs. One of the government's political failures—the National Industrial Recovery Act (NIRA) of 16 June 1933—proved to have far-reaching effects on the nation's timber resources, with ramifications for the future of timber management on the national forests. The NIRA, administered by the National Recovery Administration (NRA) under the sign of its distinctive blue eagle, was an ambitious experiment in corporate socialism, a nationwide system of governmental-industrial cooperation modeled vaguely on the mobilization economy of World War I. Each class of the country's industries was to be organized and regulated in a compact that would control the economics and mechanics of production, distribution, and marketing. The aim was to revive the industrial economy, eliminate cutthroat competition, and improve working conditions by imposing Codes of Fair Competition.

Industries that were producing natural resources were an especially challenging group. Petroleum, coal, lumber, and others had some things in common, not the least being diverse forms of ownership and regional interests. Each was sorely depressed by excess production that generally predated the depression. Each was torn by conflicts between proponents of conservation and those concerned with immediate financial survival. Each had had a brush with the issue of governmental regulation, and each included factions that favored regulation or cooperation. Each involved a lot of money, employment, and political influence.

The lumber industry was among the hardest hit of commodities suppliers, given the dearth of construction. It was to be regulated by the Lumber Code. Earle Clapp and others in the Forest Service perceived that NRA authorities could be used as a means to control destructive logging in private industry. Clapp's former assistant Ward Shepard urged the president directly to use the Lumber Code as a means of implementing important parts of the Copeland Report. Secretary of Agriculture Henry A. Wallace seconded the motion, asking Roosevelt to charge the Forest Service with administering the conservation parts of NIRA that pertained to the lumber industry.

Roosevelt was a strong believer in forestry and resource conservation, and his program of controlling all industries through central planning and cooperation reflected his attitude toward the lumber industry. He directed Wallace to tell the National Lumber Manufacturers Association, which was coordinating the drafting of the Lumber Code, that "the President asks me to tell you that he trusts any code relating to the cutting of timber will contain some definite provision for the controlling of destructive exploitation." William B. Greeley and David T. Mason responded quickly with a draft of Article X of the code, which set minimum standards of logging performance. The Forest Service countered with its own draft—strict regulations that Mason called "vicious." When the two parties arrived at a compromise, the president rejected it as being too lenient; he wanted stern controls. By 19 August 1933 he had what he wanted.

The Article X requirements, to take effect in June 1934, required operators to submit management plans to the Lumber Code authority for approval. Mason, Greeley, and other industry leaders labored long and hard to bring the entire industry under the code as soon as possible. However, on 27 May 1935 the Supreme Court ruled that the NIRA was unconstitutional, because it exceeded the authority of Congress under the Commerce Clause of the Constitution—one of seven such decisions that undercut New Deal programs.[4]

Abortive as the NIRA program had been, it helped to transform the practices of the logging industry, and it significantly advanced the cause of sustained yield as preached by Mason and, increasingly, by Greeley. It also caused the lumber industry to become more unified than it had ever been before. Industry spokesmen generally thought that the brief NIRA experience had been successful, that it had brought enduring results. Several claimed that it had ended forest devastation almost overnight, increased the cooperative spirit, and built a closer relationship between the industry and the Forest Service. Privately, however, they were just as pleased to see the Lumber Code abandoned, because it was apparent that conservation would never be successful wholly as an imposition from above. They retained the conservation principles of Article X on a voluntary basis. The National Lumber Manufacturers Association announced that it would voluntarily retain these principles, and three-quarters of its affiliates signed up. The association hired seven foresters to oversee the conservation effort.[5]

NIRA had indeed been compatible with Mason's idea of sustained yield, which he continued to urge on the Forest Service, especially in the Northwest. Greeley believed that politically, sustained yield would benefit industry, which had theretofore lacked a program of its own beyond opposing notions advanced by the Forest Service. Greeley said even more:

> I take 1933 as a turning point because that date marks the first industry-wide effort consciously to log its forests so that the land will restock with trees. . . . The timber associations took their obligation seriously. Forest-cutting rules were adopted. Foresters were employed to explain and enforce them in the woods. Meetings were held with loggers up every branch creek. It was practical education; and it took hold. . . . The groups which had sponsored the NRA lumber code voted to carry on its conservation provisions as a voluntary program, without benefit of legal sanctions. Association foresters went right on, inspecting logging operations, explaining the rules of cutting practice, and doing their best to obtain compliance. Probably two thirds of the operators conformed, at least with the spirit of the forest code.[6]

Despite Greeley's rosy picture and despite the fact that the industry did greatly reform itself, the experience revived the crusade for federal regulation of the timber industry. That, in turn, became caught up in Secretary of the Interior Harold L. Ickes's attempt in the late 1930s to get the national forests transferred to his own department. Meanwhile, Mason and his associates gained a significant victory in 1937, when Congress authorized the establishment of cooperative sustained-yield units involving private and Department of the Interior lands in the Northwest. The Forest Service resisted the idea of cooperative sustained yield in favor of the federal regulation of private lands; its own timber sales were so low that the only hope it had of controlling harvesting practices was on the lands of others. Forest Service leaders engineered a call by the Natural Resources Board for the Copeland Report program, emphasizing the public acquisition of mismanaged lands and the regulation of others. Although the board generally aimed at a holistic scheme for all natural resources, it acquiesced in the Forest Service's assumption that the production of timber was the dominant purpose of the nation's forests.[7]

The regulation crusade was doomed from the outset, and not only for the obvious reasons that it was unacceptable, unconstitutional, or both. Even as spokesmen for industry were calling for the establishment of cooperative sustained-yield units, some in the Forest Service echoed them, or appeared to. Moreover, the Forest Survey produced figures that showed, especially in the Northwest, that there was more timber than had been believed; that even the boom harvests of the 1920s had not "drained" the forest resources, because reproduction had kept pace; that industrial logging was not as uniformly destructive as previously believed; and that private operators were in fact being increasingly careful about their harvesting methods.[8]

Timber remained uppermost in every Forest Service calculation, or so it appeared. During a debate on whether the Forest Survey should account for timber on Primitive Areas (administratively reserved from development), the Pacific Northwest regional office pointed out that "every Primitive Area has of course been selected with the idea of using lands of little or no value for timber or other commercial uses, unless it might be grazing."

The Coconino National Forest, tracing the decline of its timber harvests from a high of 31,769 MBF in 1929 to 388 MBF in 1933, sought to recover its old market. Perceiving that its former policy of assigning territories to several firms had contributed to the duplication of facilities and to reduced purchasing power, the Coconino decided to revise its management plan to encourage "the far more desirable and almost inevitable reorganization of the industries." In other words, it would throw timber on the market in such a way that only one firm would be left in the Flagstaff working circle. To meet the goal of timber sales, the old watchwords about stability and the prevention of monopoly went out the window, with the hearty approval of the Washington office.[9]

The frustrating attempts to restore a high volume of sales of national-forest timber appear more obsessive than they probably were. Effort was bound to exceed reward when the market was so depressed. That it was made at all may have been a nostalgic reference to the 1920s, when high sales figures compensated for earlier frustrations. In addition, during the 1920s the Forest Service was coincidentally blessed by high morale and was relatively free of bloody political conflicts. The circumstances were reversed in the early 1930s: sales were down, and conflicts were up.

Sales were also weapons in the tug of war with private industry, which, because of overproduction, was extremely vulnerable on the local level in many parts of the country. On 15 September 1934, during the life of NRA, the Forest Service issued a policy memorandum on sales, which was restated and amplified in 1936, after NRA fell. This memorandum asserted: "National Forest timber sales will be used to the fullest possible extent consistent with the law and any inescapable practical considerations to promote or enforce sustained yield, or, where that is impossible, at least crude forestry on intermingled or adjacent private lands. National Forest timber will not be sold to applicants who are operators on intermingled or adjacent private land . . . who refuse to practice sustained yield.''

In addition, purchasers of national-forest timber ''must agree'' to practice a minimum degree of forestry on their own lands. They were limited to national-forest areas no larger than the private areas on which they would ''agree to forestry practices.'' Furthermore, ''If found legal [the Forest Service would specifically reject any bid unless the purchasing landowner] agrees to practice the stated forestry measures on his own land.''[10]

Timber sales, then, could be the next best thing to regulation. Whatever the sentiments, timber sales assumed a general importance wherever they were conceivably possible, conditions of the market notwithstanding.

FIRST TEST OF MULTIPLE USE

There were other pressures on the national forests, however. Recreational visits were estimated at 13 million per year by 1935. Many recreationists were sportsmen, hunting wildlife; others went to the forests for things that depended upon timber only as scenery. Watershed protection had been the original purpose of many national forests, and it was supposed to be important in the management of all of them. Lastly, the heaviest users of national forests since the turn of the century had been graziers, not timbermen. Each interest had caused the assignment of Forest Service personnel to plan and manage it, and each accordingly had its spokesmen in the agency. It might seem that all the talk was about timber, but it was not. A dividing of the collective agency mind was beginning.

To reconcile what could be conflicting interests, Chief Ferdinand A. Silcox proclaimed in 1935 that the national forests ''must be managed on the principle of multiple-use. . . . This multiple-use principle is not new. It has been applied to the national forests for more than 30 years.'' In a sense he was correct, for various uses had been made of the forests for decades. It was only now becoming apparent that some kind of coherent managerial philosophy might be required when differing uses (or interests) came into conflict over the same ground. The concept was vague in 1935, however, and at its root lay a grasp for an argument that would prevent recreational lands from being transferred to the National Park Service. In promoting multiple use, the Forest Service was really saying that it could manage recreation as well as timber, and both of them together.[11]

To name a concept is not necessarily to formulate it, and the Forest Service's use of the terms *multiple use* and *sustained yield* in the mid 1930s betrayed a lingering commitment to timber as the lifeblood of the nation and the first purpose of the national forests. The timber famine might have gone, but it was not forgotten.

Pronouncements made by the Pacific Northwest Region were revealing. Early in 1935, Regional Forester C. J. Buck announced a grand vision of "New Deal Forestry," to preserve Oregon's "principal manufacturing industry, and [involving the] utilization of the natural product of nearly half of the state's total land area." The plan was founded upon:

1. Adoption of state and federal policies "favorable to sustained yield"
2. Intelligent land classification
3. Increased protection against fire
4. Adjustment of taxes on timber
5. Provisions for federal credits for cooperative sustained-yield units
6. The formation of sustained-yield units after the fashion of "drainage [soil and water conservation] districts"
7. Increased federal and state land acquisition "to complete sustained yield units where private ownership is not adequate"
8. "Equitable compensation" to counties when such lands were removed from tax rolls
9. Coordinated research under federal auspices
10. Protection of watersheds
11. Development of the maximum use of recreational opportunities
12. Protection of game and wildlife
13. Securing permanence of the range resources[12]

What the Forest Service meant by *sustained yield* and *multiple use* became very clear in further news releases that grew out of the region's surveys. Where there was timber, "sustained yield" management was the way to go. Where there was no timber, "multiple use" was the prescription. In Oregon's Umatilla County, for instance, the timber industry was "comparatively unimportant." Therefore, "the principle of multiple use planning, as practiced on the national forests, is particularly important in a county much as Umatilla, according to the foresters. Such a plan seeks a balanced use and continuous production of the various forest contributions such as lumber, forage, watershed protection, and recreation. This principle of multiple use planning should be extended to all forest lands in the country, if the best public interest is to be served, it is said."[13]

In contrast, Multnomah County processed a lot of timber, despite the depression in the timber industry: "The survey indicates that, because of its virgin timber supply, Oregon is one of the few states in the Union where a sustained timber yield program can be put into operation without curtailing industrial activity pending the time when new forest crops will be available for cutting."[14]

This attitude—that timber claimed first priority, that multiple use should be instituted only where it would not interfere with potential timber harvests—undermined multiple use as an argument that the Forest Service could manage recreational values as creditably as could the National Park Service. It thereby brought defeat to the Forest Service in one of the bitterest conflicts that the two agencies ever waged over a piece of ground—Olympic National Park in Washington.

Olympic National Monument had been established in 1909; in 1915 President Wilson had reduced it so as to restrict it to higher ground in the middle of the Olympic National Forest. It was administered by the Forest Service as part of the national forest until the Park System was reorganized in 1933. At about the same time, public controversy arose over Forest Service efforts to reduce the badly overpopulated herds of Roosevelt elk in the area. Conservationists never approved the reduction of the national monument, and in the 1930s they launched a strong campaign for an Olympic National Park—to be expanded from the Monument at the expense of the national forest. The Forest Service resisted such a transfer as a matter of course, but its single-minded focus on timber undermined its case from the outset. This ultimately antagonized President Roosevelt and led to the establishment of the national park in 1938.

In 1935, Supervisor H. L. Plumb of the Olympic National Forest advocated publicly that "the entire area be administered by the forestry service with the timber cut on a selective logging, sustained yield basis, with care taken to preserve the recreational features and wild life." Not content merely to advocate timber cutting in the area, Plumb implied that the national monument should be removed from the Park System and returned to the national forest. Proponents of the national park were galvanized.[15]

Both camps soon were at it tooth and nail. Supporters of logging and those of the park were about evenly divided in the region, the latter looking to tourism as a more stable local industry than timber. The former said the park proposal was "carrying the scenic fad too far." One of the latter examined logging in the vicinity and said that loggers "talk of selective logging but that's all rubbish. About the only selective logging the peninsula knows is that loggers select the trees they want—and cut them."

Park Service Director Arno B. Cammerer told Congress that the "issue involved is between the Forest Service and lumbering, on one side, and conservation on the other." The Forest Service's Leon F. Kneipp responded that sustained yield and fire protection, under the

Forest Service, would make the area "permanent." "We do not need protection from timber cutting," he said; "we need protection from high speed highways, big busses, hotels, and so on."

By 1936 the Forest Service apparently perceived that it had weakened its case by trying to expand timber harvests in the disputed area. It established a Primitive Area of almost 239,000 acres. According to the regional forester, "recreation and wilderness . . . have been achieved without serious economic injury to the state." Interior Secretary Ickes countered by saying that selective logging would be practical within his proposed park, cutting timber in strips on a one-hundred-year rotation. Agriculture Secretary Wallace and the Forest Service lost the case by implying that there were plans for timber harvests, power developments, manganese mining, and other developments that were intolerable in a national park. Moreover, when President Roosevelt toured the area, he was outraged by scenes of clearcut logging. What he saw was on private lands abutting on the national forest, but he blamed the Forest Service.

So much for multiple use, at least in its first real test. The president and Congress decided that the wildlife and other natural values required complete protection; they apparently concluded that the Forest Service intended to make these things secondary to economic development. Olympic National Park was established, and Olympic National Forest became much smaller.[16]

The Forest Service became a victim of its own loose talk. There is little doubt that timber was uppermost in the Service's mind and that in the Northwest, it as much as said that multiple use would be implemented only where it did not conflict with timber. However, the Service was rather more broad-minded than it appeared. The fact was that the lumber market continued to be depressed, and talk that made people expect large-scale timber harvests on the national forests remained just that—talk. Fueled by CCC manpower and New Deal money and relieved of a burdensome sales volume, the Service was involved deeply in studies intended to refine timber-management methods. Minor subjects of attention, destined to grow in importance, included studies of insect control, the chemical control of vegetation in firebreaks, techniques for thinning and for other "timber stand improvement," and the use of trucks in logging. The depressed timber industry aside, further studies suggested that fires were a greater cause of timber depletion than were logging practices.[17]

In the Northwest, where many of those who feared the threat of logging in Olympic National Park probably remembered both the ghastly harvests of the cut-out-and-get-out days on private lands and similar railroad logging on national forests, new information was destined to change Forest Service practices if large sales ever resumed. The general adoption of motorized equipment, combined with changing market demands for sawtimber and pulpwood, made selective logging increasingly economical. Even the tedious job of marking for selective cutting was expedited when a Forest Service man hit upon the idea of making paint guns out of filling-station oil cans. The Forest Service helped the Edward Hines Lumber Company develop an annual selection harvesting system for its lands in eastern Oregon. Most interesting, however, was the finding that selective harvesting proved most advantageous in species that had long been considered candidates for clearcutting—most notably ponderosa pine.[18]

Sentiment for David Mason's version of sustained yield grew in the Northwest, especially during 1936. Legislation was introduced in the Congress, and its sponsors toured the region to stump for support. The Washington State Planning Council resolved in their favor: in the establishment of sustained-yield units, it urged that harvests on national forests "be limited to an amount which can be permanently sustained without reducing the volume of forest control." The Forest Service, meanwhile, had established the model Monte Cristo Ranger District, near Verlot, Washington, to prove the worth of sustained yield and selective logging. Previous utterances in the region notwithstanding, there the Service promised multiple use, with timber not necessarily being first among equals: "The principal objectives of the service in this district will be forest protection, recreational stimulation, timber management, and wild life management, and especially, under a program of multiple use, to coordinate the recreational and timber management activities so they will not conflict."[19]

A NASTY FIGHT OVER REGULATION

The First New Deal came to its (finally) triumphant conclusion with Roosevelt's reelection in 1936. Leaders of the Forest Service, like many others in the country, interpreted the president's victory as general approval of the central economic management he had tried to impose on the nation. Comparing events in Europe with those in the United States as they affected forestry, Chief Ferdinand Silcox said that both exhibited a common trend in favor of centralized control. In his opinion, "the United States must sooner or later accept regulatory control." Earle Clapp agreed, saying: "The recent election indicated a mandate for social control in economic fields."[20] The Forest Service accordingly prepared to do righteous battle.

At a meeting of regional foresters and directors of experiment stations late in 1936, Silcox remarked "that the Washington Office would naturally take the lead in establishing Forest Service policies; that the Regions would of course be definitely in the picture and that when policies were adopted all forest officers were expected to accept them like good soldiers and see that efficient application was made.''[21] He wanted to have his agency present a solid front, because he was in the midst of a renewed campaign for federal regulation of private-timber harvesting practices. It all turned on two interpretations of the term *sustained yield*.

To many people in the timber industry, sustained yield meant the sustenance of the industry—continuing controlled supplies of raw material to stabilize local economies over the long term. Since that would require nondestructive harvesting methods, it would also serve the end of sustaining the forests. Industry spokesmen claimed that destructive logging practices were nearly extinct. Conservation goals, they averred, could best be met by establishing cooperative sustained-yield units involving all the forest lands, public and private, in a given geographic area, with production coordinated in ways evocative of the NIRA system. Some in the Forest Service agreed and favored the establishment of sustained-yield units. But the official positions of the agency were that cooperation had not worked and that federal regulation of private industry was necessary if the Service were to achieve its version of sustained yield—continued productivity of the nation's forests by balancing harvests and reproduction.[22]

Silcox needed a unified Forest Service position because foresters at large were no longer behind the cause of regulation. Many of the members of the Society of American Foresters were employees of industry—they represented reformed industrial practices that Silcox and his allies refused to acknowledge. Silcox claimed that industry had nowhere attempted to reforest cutover lands, millions of acres of which were being abandoned to delinquent tax rolls or to the Forest Service. Although he moderated his regulatory program by proposing decentralized administration, to an important part of the Society of American Foresters his ideas were socialistic.

During the late 1930s the fight over regulation was one of the nastiest ever waged, with extremists on both sides comparing their opponents to the various totalitarians ruling in Europe. It shared the political spotlight with equally vicious struggles over the control of electrical power, petroleum, coal, and other commodities—which together helped to kill all but two of the proposed federal regional resource developments (the Tennessee Valley Authority and the Bonneville Power Administration) and made hopeless the kind of comprehensive governance of all natural resources that the Natural Resources Board espoused. Regulation was a doomed cause, because there was a growing reaction in the country against tendencies in the New Deal that were regarded as socialistic and out of keeping with American life ways. Extremism and recriminations merely served to seal the doom.

Even President Roosevelt was cautious. Though he agreed with Silcox in principle, he took the moderate approach of seeking an investigation in Congress, which at his request established the Joint Committee on Forestry in June 1938. The committee received from the Forest Service a tendentious call for regulation, which promised an end to all social evils and was leavened with denigration of the timber industry. The Service tried to dominate the committee hearings, thus bringing upon itself strong reactions from industrialists, with the Society of American Foresters and the American Forestry Association equivocating on the sidelines.

While Greeley was attempting to develop a program for the industry—founded on cooperation and regulation at the state level—Silcox and Clapp were papering the Forest Service with confidential letters, telling field personnel to support the crusade. In one message, Clapp said: "We should not allow ourselves to be deterred by fear of criticism by timberland owners or reactionaries whose point of view is frequently selfish and contrary to public interest." He told the men in the field to follow orders. That unfortunate letter fell into the hands of H. H. Chapman, an opponent of the Forest Service's position, and produced a scandal. Another academic forester called Clapp's directive to the Forest Service "socialism reduced to dictatorship."[23]

Early in 1940 the Forest Service forwarded a long set of recommendations to the Joint Committee on Forestry. Most interesting was the agency's endorsement of cooperative public/private sustained-yield units. However, the proposals emphasized regulation "as part and parcel of cooperation," because industry alone could not be trusted to protect the public interest. To administer a regulatory program, the Forest Service wanted itself to be in charge, but its fall-back position would have passed that duty to the states, with federal intervention when states were negligent. Least desirable to the Forest Service was increased cooperation to encourage an end to destructive logging.

When the Joint Committee issued its report in 1941, it only superficially supported the Forest Service's position by calling for regulation—but by the states, not by the federal government. To Roosevelt the proposal was "pitifully weak." A "Forestry Omnibus bill" appeared in the fall of 1941, but without support from the Forest Service and with opposition from industry, it vanished in committee. The futile regulation crusade had done little more than inflame passions and leave a bad taste in everyone's mouth. The Forest Service appeared decidedly disinclined to cooperate with industry except on the Service's own terms.

The centralization in the Forest Service administrative system that accompanied the regulation crusade met growing resistance within the agency. Regional Forester Allen S. Peck of Region 2 thought that the growing numbers of inspections during peak work periods added to workloads at the local level. The aggregation of staffs in the Washington and regional offices, he believed, inevitably left the district ranger in an "unfortunate position."[24] The ranger's workload was not eased by the fact that the timber appraisal manual had last been revised in 1922 and did not mention such things as mechanical equipment.[25]

OTHER BATTLES OVER HARVESTING

The cutting of timber in the Northwest remained a subject of debate. There seemed to be no satisfactory answer to the question of how—or how many—seed trees should be left in a heavy cutting area or how to protect them from destruction by winds. The trend seemed to be in favor of less clearcutting and more judicious selection. A large sale of ponderosa pine in eastern Oregon called for the marking of only 40 percent of the timber, a departure from the heavier cutting of the past. When the president and his party toured the Northwest, they objected to the clearcutting that they saw on national forests in western Oregon and Washington. Regional Forester Buck explained that it had occurred where private lands were intermingled with federal lands, so as to protect the latter from fire or blowdown after private logging. Strips of timber had been left along highways, but the trees had blown down in winter storms.[26]

The Forest Service's objectives of sustained-yield harvesting could not be met simply, and factors that had not occurred to the federal men were raised by the industry. At a meeting late in 1937, timber operators of the Northwest suggested that the "economic selection" then being promoted by the Forest Service might flood the market with high-quality material and cause further disruption in the industry. On the other hand, the Forest Service's projected shorter rotation periods and future harvest of young second-growth timber caused its own Benson Paul to question where clear (knot-free) lumber, currently a mature-timber product, would come from. Meanwhile, the Forest Service announced an increased program of exchanges of government timber for private land and timber, to "eliminate land stripping by private operators."[27]

"The growing of timber is now coming to be recognized as but another form of agriculture," said the supervisor of the Coconino

National Forest in 1939. There, "to assure continuous crops of timber," the Forest Service engaged only in selective cutting, figuring that it took sixty years to raise a "crop." That was accomplished "by harvesting only the ripe timber; reserving the young, thrifty, growing trees; removing diseased, defective, and otherwise undesirable trees; making thinnings in the young trees to secure accelerated growth, and insuring reproduction when not already present by the retention of seed trees." That was a sophisticated pronouncement from a region where most Forest Service people were, according to Alva A. Simpson, "the cowboy type" and where forestry specialists had only recently arrived in substantial numbers.[28]

Such statements were typical of the Forest Service's efforts to sell its versions of sustained-yield harvesting to industry. Most of the attention focused on the Northwest, where most of the timber was, where the industry was most prominent, and where sales of national-forest timber were at higher levels than elsewhere. To be sure, the Service was pleased to observe sustained-yield harvesting on industrial lands elsewhere, as in the South; but it was the Northwest that drew the most attention and where the campaign for regulation was taken to the "enemy's" greatest stronghold. The newspapers were full of it—articles extolling instances of industrial support of sustained yield (but uniformly its own version, not the Forest Service's) and others citing the statements of Forest Service leaders, predicting disaster for the region if the Service did not have its way. When Clapp or Silcox came to town, the timber famine always reared its ugly head.

Regional Forest Service people seemed to get along better with the industry than did the crusaders from Washington. The most positive event of the period was the agreement between the Forest Service and Shevlin-Hixon Lumber Company of Bend, Oregon, to allow the Service to mark all timber, private and public, in its harvest program. The company was not persuaded that the Forest Service's system would pay off, but it was willing to give it a try as long as both ownerships were involved. As it affected the Forest Service, this was the first small triumph for David T. Mason's version of sustained yield. Moreover, it was a means by which to expand national-forest acreage, as the government acquired company land that had been cut over under Forest Service standards in exchange for cutting rights on national forest land; thousands of acres were acquired by such means. The Weyerhaeuser Company, meanwhile, announced the establishment of a 130,000-acre tree farm in Washington, "the first of its kind ever to be attempted by a private concern." One firm, at least, planned to go its own way.[29]

While the hierarchy beat the bushes to promote regulation, Forest Service people at the working level still had jobs to do. The promotional dimension was expressed at the local level as well, however. The Forest Service's language increasingly emphasized selective harvesting and deemphasized clearcutting in the late 1930s and early 1940s, although there was more talk than action in regard to selection. Both ponderosa pine and Douglas-fir were important targets of pronouncements on selective cutting—and of the regulation crusade. The public was told that the selective cutting of lodgepole-pine forests caused more snow in the water-starved West. On the other hand, some researchers feared that selective cutting might promote destructive insects, "which raises questions as to the economic consequences of leaving certain kinds of trees in the reserve stand." Insects, in any case, were to be controlled with common sense and appropriate restraint. Meanwhile, the Modoc National Forest developed a management plan for the Big Valley Working Circle; the plan was designed to stabilize local industries, although some reviewers thought it gave insufficient consideration to recreation, esthetics, and watersheds.[30]

THE MORALE OF THE FOREST SERVICE

The burden of routine work could not erase the fact that the Forest Service was in a despondent mood. In its crusade for regulation, the Washington office asked field people to promote something that many of them found repugnant. It also made living with neighbors difficult in areas where disputes were the most bitter. At their conference in February 1940, regional foresters learned that morale was low and that the Forest Service was failing to attract young foresters, who found private industry a better place in which to work. Furthermore, the crusade had demanded public relations, and "the average technical forester dislikes this work, and he is not too good in it." The Washington office proposed to hire "a limited number of specialists" to do things that foresters were not qualified to do and to start a propaganda campaign to entice "the best men" into government instead of industry.[31]

When Earle Clapp toured the West, he found the Forest Service beset by difficulties, foremost among them being low spirits:

> *One group of problems* included the existing morale in the Forest Service, and baffled feeling of many men because they do not know what Forest Service objectives are, the feeling of uneasiness or even hopelessness because of inadequate legislation and funds for badly needed work, the belief that the Service is continually on the defensive, the damper on enthusiasm and creative effort caused by the threat of reorganization, the tendency throughout the Service toward provincialism and a narrow conception of individual jobs and of Service responsibilities, etc., etc.

Forest Service people, he said, frequently worked at cross purposes, enough to "prevent real progress." His solution was an "educational campaign" to "stimulate aggressive public action" on the issue of regulation. That, he believed, would restore "morale and fighting spirit." But it was just such a crusade that had contributed to low morale among the Forest Service's own people and had caused others to go to such extremes as to say that "the Forest Service has found itself in a position where, in its anxiety and desire to dominate the forest products industries, it must ignore, in fact it must sabotage private forestry efforts."[32] That charge may not have been true, but it did reflect the way the agency too often let itself be perceived by others with whom it should have had only honest differences of opinion.

ANOTHER WAR, ANOTHER CRUSADE

The regulation campaign, of course, was a fizzle by early 1941. The Forest Service was about to join another crusade that would find it in harmony with an equally mobilized nation. Even the long-depressed lumber industry had fallen on boom times. Mills in the Northwest turned out lumber at 90.6 percent of capacity in 1940—up from 75 percent in 1939 and 45 percent in 1938.[33] What was happening was the advent of another world war. Beginning with appropriations in August 1937 and June 1938, the War Department resumed a permanent army construction program that had been interrupted in 1933. In the autumn of 1938 the government embarked on a limited national-defense expansion program, including military and industrial construction. By 1940 the program resembled mobilization, and after the Japanese attack on Pearl Harbor in December 1941, the War Department's construction program approached $12 billion.

Expansion and mobilization construction was a shot in the arm for the construction and building-materials industries, as defense procurement was for all industries. Lumber surpluses were absorbed fairly rapidly, and by 1940 the industry had emerged from depression and was straining to fill orders. The demand was so great, in fact, that lumber shortages were a defense construction problem into 1943, when the construction program fell off.[34]

Because of the accelerated demand for timber for defense purposes, demand for national-forest timber increased. For the first time, national forests provided a significant part of the nation's current lumber supply. In 1930, when the domestic production of lumber totaled 29.4 billion board feet, national forests sold only 1.048 billion board feet of timber. When the lumber market hit bottom in 1933, total production was 17.2 billion board feet, while sales of national-forest timber amounted to 389 million (0.389 billion) board feet. National production of lumber finally reached 31.2 billion board feet in 1940, when the Forest Service sold 1.371 billion feet of timber, a measurable increase in its share of the total when compared with 1930. The national production of lumber peaked at 36.5 and 36.3 billion feet, respectively, in 1941 and 1942, then fell off to 28.1 billion feet in 1945. National-forest sales, however, did not fall off; they increased during the war: 1.552 billion board feet in 1941; 1.560 in 1942; 1.864 in 1943; 2.840 in 1944; and 2.732 in 1945. By the end of the war the share of the nation's wood production that was reflected in sales of national-forest timber had increased by 250 percent, with national forests supplying roughly 10 percent of the nation's lumber supply.[35]

The Forest Service had long predicted a time when private stocks would not be able to meet the nation's timber demands and the national forests would have to fill the shortfall. It had not predicted that the time would come so soon, nor in such a transitional fashion. Harbingers of the timber famine had said that America's clamor for wood would come in a calamity. A world war was not the kind of calamity they had expected. Everything about the war and mobilization was unprecedented, not least the scale of demands for forest products. If the harbingers of disaster had been correct, neither the forests nor the nation should have survived the ordeal. Both did, however, and along with them the National Forest System, itself as altered by the great test as was the United States.

The Forest Service and the timber industry entered World War II in an uneasy truce. The national forests had long been regarded mostly as reserve stocks to meet future shortages. Because of the need to withhold federal timber from the market to avoid further disruption of private industry, the federal foresters had not instituted intensive management aimed at high production. When mobilization demands lifted the fear of competition with private timber, the industry and its dependents were suspicious of Forest Service intentions. Said one editor in the Northwest, in a not uncommon fit of innuendo: "Like so many things that are done under the cover of a popular phrase [namely, conservation], there is another side to this government venture into timber. And if that other side is even half what numerous lumbermen describe it as being, then the fine old institution known as the U.S. Forest Service begins to take on a new appearance—one that smells, in fact."[36]

Public feuds over such issues as regulation were mostly put aside for the duration of the national emergency, however. Both the Forest Service and industry were overwhelmed with demands for timber. The Forest Service stepped up its sales and harvests, while 90 percent of the billions of board feet required for the war effort were supplied from private stocks. Wood was required for innumerable uses; Girard of the

Forest Survey, for instance, spent most of the war lining up stocks of aircraft materials. The work went forward in a system of national priorities and allocations involving all parts of the public and private sectors, boosted significantly, especially in the South, by the Timber Production War Project. The Forest Service, however, did its work with limited manpower—about two thousand Service people had joined the armed forces by the end of 1942.[37]

A great deal of attention was devoted to fire protection, as much for immediate security reasons as for the long-term protection of timber; the absence of CCC manpower (the CCC was abolished in 1942) was partly made up by the use of prisoners of war, reserve troops, and an expanded smokejumper (parachute-borne fire fighter) program, while a series of "good" fire years helped. Prisoners were also used in logging operations. The wartime salvage of 100 million board feet of storm-damaged timber in Texas was handled mostly by private industry. In the Forest Service, however, an inspection report in 1945 said that "as a matter of war time expediency many men have been jumped into jobs for which they have not been fully trained," with the result that "there is some good timber here that may be spoiled." Private-sector proponents of sustained yield were worried that the large harvests of the war, if not carefully directed, would invite trouble in the future. Said one, "It is possible, nay probable, that the need for lumber for war purposes may be used as an excuse to postpone more general adoption of this selective logging system in the remaining timbered areas of the Pacific Northwest, but if this excuse is accepted, then it is going to be too bad for all, except perhaps the few who are intent only on making the most money possible out of the forests in the shortest possible time."[38]

The forest-products industries were profoundly affected by the wartime experience. Even before the war a number of the larger firms had given up faith in David Mason's version of cooperative sustained yield and had embarked on ambitious programs to acquire and permanently manage their own timber; one firm alone acquired 400,000 acres on the Olympic Peninsula to ensure its own pulpwood supplies, while several in the South made similar moves.[39] Sustained-yield logging became general on large private ownerships, while improved fire control brightened the prospects for future growth on cutover lands.

THE WARS OF PEACETIME

Promising as those developments were, they were not enough. The war accelerated the extraction of mature timber stocks, and when a postwar housing boom appeared likely, demand for public timber increased. Private operators, whether owners of timber or not, required increased access to national-forest timber in order to meet current demands while reforestation was proceeding on cutover lands. After the war, industry spokesmen clamored increasingly for access to the federal resource, lest they be forced out of business for want of supply. Some even called for decreases in the National Forest System, which was regarded as a "hoarding" of valuable material.

By the end of the war it was apparent that the national forests were about to play a larger role in the nation's timber economy. The question was whether the national forests or the Forest Service were ready for the new circumstances. E. E. Carter, who retired in 1945 after more than two decades at the top of the Forest Service's timber programs, believed that change was natural and that the Forest Service was ready. "Future foresters," he said, "will deride our ideas, efforts, and accomplishments. That will be as it should be. Let them do better, starting where we enabled them to start."[40] As for the forests, Carter told the regional forester in the Pacific Northwest that "the contrast between present development of the use of that timber resource and conditions when I first visited these Forests in 1906 points strongly to the future intensification of timber management on these areas which have heretofore been regarded as storehouses, to be drawn in the indefinite future."[41]

In 1944, legislation authorizing cooperative sustained-yield units promised to integrate national-forest timber management with that of neighboring lands. However, cooperation between the Forest Service and industry became increasingly difficult, as the two became embroiled in another controversy—the worst yet—over the regulation of private lands.[42]

Earle Clapp, acting chief of the Forest Service from 1939 to 1943, did not give up the regulation crusade altogether after the fizzle of 1941. His efforts to take advantage of the war as a means to his end came to little during 1942, however. He was replaced by Lyle F. Watts in January 1943. Watts was one of a series of Forest Service chiefs who had served important stages of their careers in the timber-dominated Pacific Northwest. He was oriented toward timber, and he believed in the federal regulation of private industry. He renewed the crusade at the end of the war, thus setting off a ferocious struggle.

Industry was ready this time, and it effectively countered the Forest Service agenda with one of its own. It made its muscle felt politically, and in the course of things, it exerted its own influence on the Forest Service's budget. Watts's efforts to increase appropriations under the Clarke-McNary Act to around $40 million a year raised the first alarms. The industry was afraid, with some reason, that the Forest Service would use an expanded budget to win state support for regulation. The fears were compounded when official instructions came to public notice, reserving to the Washington office the power to recommend appointments to state positions. Industry support for the Clarke-McNary program softened after World War II, because the fire-protection benefits of the act were not regarded as worth expanding into a complete regulatory system; desires to trim the federal budget and to reduce the wartime debt were also involved. As related to the budget for the national forests, industrialists supported funding for timber sales, but they stiffly fought appropriations to expand the federal forests.

Watts further eroded relations with the industry by advancing renewed plans for the governmental regulation of private operations and lands. His program called for regulation by the states, but it left two big whips in federal hands—the power to step in when the states neglected their roles and the authority to acquire private lands for the National Forest System when the Forest Service thought this to be in the public interest. Ownership and management of timber by the federal government would be a tool of regulation. His plan took the form of draft legislation in 1949, but thanks to industry opposition, the measure never reached a vote. Still, the open feud between the Forest Service and the industry continued for another three years.

Meanwhile, industry had already detected what some of its spokesmen claimed was the manipulation of national-forest timber management in order to gain control of private industry. That was the "hoarding" controversy of 1946. Watts claimed, in his annual report that year, that the national timber supply was dwindling at a rate of 18.6 billion board feet per year, attributable mostly to poor logging operations on private lands and making public regulation necessary. The American Forestry Association, meanwhile, issued its own review of the national timber picture. Although it agreed that depletion had occurred, it put the blame on public agencies. Of the nation's timber, 40 percent was publicly owned but contributed only 9 percent of the annual harvest. AFA wanted public aid for industry but wanted regulation at the state level only.[43]

The AFA report, as well as others in industry, acknowledged that public timber had long been held from the market at the request of industry, to reduce competition. As Senator Guy Cordon of Oregon understood it, "this was done originally to help private owners of timber lands reap the harvest from their timber."[44] Circumstances had changed, however. One major part of the "reconversion" from a wartime to a peacetime economy was a large, federally supported housing program for veterans, which added to a postwar construction boom caused by sudden release of demand and wherewithal, long pent up by depression and war. All the new construction would require a lot of materials, and housing officials pestered the Forest Service to expedite the production of lumber.

Spokesmen for the timber industry, having difficulty meeting new orders from their own sources, accused the Forest Service of hoarding public timber. According to H. J. Cox of the Willamette Valley Lumbermen's Association, the Forest Service did not "intend to contribute to the economic welfare of our commonwealth unless compelled to do so." He said the Service was deliberately holding onto its own stocks until private timber had been depleted by the current high demand. Then, "bureaucratic control from Washington D. C. can dictate the policy of our major industry as well as govern the destiny of those directly and indirectly dependent on such industry for livelihood." Regional Forester Horace J. Andrews responded from Portland with a flat denial, saying that his region had sold 19.5 billion board feet of timber since 1909 and that timber was available for competitive sale, but only for sustained-yield cutting.[45]

Cox was more extreme than most, but much of the industry took up the charge of hoarding. For a while, it was effective as a counter to the regulation campaign; the president of the West Coast Lumbermen's Association suggested that the "tremendous demand for housing might postpone establishment of sustained-yield lumber cutting" on private lands. The implication was that hoarding by the Forest Service was undermining industry's efforts to practice conservation. The fuss about hoarding had also presented the industry with an alibi for any shortfalls in production. However, when it was pointed out that opening up the national forests would require public timber-access roads that would be open to small operators as well as large ones, who then controlled most private logging roads, the whole controversy vanished overnight. Instead, the National Lumber Manufacturers Association called for an end to the expansion of public ownership of timber and even suggested a reduction of public ownership.[46]

THE ACCESS PROBLEM

To a certain extent, critics of the Forest Service had a point. The agency was hoarding timber, not deliberately, but because of inadequate access. Construction of timber-access roads was generally the responsibility of purchasers of national-forest timber, the costs and value to the government being figured into appraisals. Since trucks had generally replaced railroad logging, roads were an important consideration.

The federal government had paid for road development to expedite production during the war, and at the end of the war the question was up in the air. It was aggravated by the fact that the wartime demand for timber, together with the fiscal and physical obstacles to road construction, had caused overexploitation of accessible tracts, while most timber in western national forests remained inaccessible. E. E. Carter said about the future in 1945:

> Purchasers cannot be expected to expend the necessary money to build these main haul roads on standards satisfactory for permanent forest management; but can be required to build extensions or side roads into their operating areas. Where we plan for this in advance and insist on the delivery of the roads as allowed for in the appraisals the long term management possibilities of the National Forests are bettered and usually the purchasers profit also through lowered operating costs from decent grades and sufficient surfacing.[47]

When the War Production Board asked for continued federal construction of access roads during the ''reconversion'' period, the Forest Service drew up a list of projects early in 1945. It declared that there was an ''urgent need for main haul roads to be constructed at public expense, if companies cutting out are to be relocated and lumber production maintained.''[48]

Carter was a strong proponent of the public construction of main-haul roads, leaving only spur roads to the purchasers. He was concerned not only with the long-term management needs of the Forest Service but also with fairness to small and large purchasers. Before the House Appropriations Committee in October 1945, he emphasized two considerations: first, higher stumpage prices would repay the costs; and second, public access roads would allow small operators to compete fairly with large firms, which were the only ones that had the wherewithal to build their own roads.[49]

Also affecting access roads was the increasing size of trucks. Late in 1945 the Forest Service considered adopting road and bridge standards to accommodate trucks with bunks that were twelve feet wide, which could significantly reduce hauling costs. According to Carter's successor, Ira J. Mason, a Forest Service construction program would be necessary: ''In so far as forest highways are concerned, it is evident that both the Public Roads Administration and the State Highway people will be quite reluctant to agree to the increased expense of accommodating these oversize trucks.''[50]

The Forest Service prepared a $10.5-million road-construction program. Mason told the National Housing Agency that this would permit increasing the annual production of national-forest timber by 700 million board feet by 1948. C. M. Granger passed the same advice to the Office of War Mobilization and Reconversion. The program for fiscal 1947 was to be spread among five regions, with the Pacific Northwest (R-6) getting one-third.[51]

Mason believed a major road-construction program would answer many problems. He avowed in July 1946 that without ''an adequate access road system a rate of cutting of about 5 billion feet will be close to the maximum allowable for the areas which probably can be opened up through operator built roads.'' In the absence of his road system, the Forest Service would continue to sell only the most accessible timber, ''without regard to whether the cutting should be done in less accessible portions of the working circle and without regard to balancing of summer and winter operating areas.'' The ''greatest gain'' from a complete road system, he promised, would be improvement in the quality of national-forest management. ''It will change it from a system of opportunistic perimeter cutting into one where the time and place of cutting is adjusted to stand conditions. It will shorten cutting cycles and speed up greatly the time when other potential improvements in the degree of utilization can be brought into effect.''

Mason estimated that there were 518 billion feet of sawtimber on the national forests, only about a third of which was accessible ''or need not be developed within the next twenty years.'' About a quarter of the remainder could be reached by operator-built roads ''without sacrifice in timber policy standards and without establishing undesirable monopolies.'' That left 260 billion feet inaccessible. To reach it, he proposed a $260-million program, extending over thirteen years, to provide a total of 26,000 miles of roads.[52]

The pending Housing Subsidy Act earmarked $15 million to develop access to publicly owned timber. The housing expeditor, however, demanded quarterly reports on the volume of timber that was produced as a result of the construction of each access road. Meanwhile, H. E. Ochsner of Region 1 reported that small operators in his Region, unlike the larger operators in the Northwest, were not financially able to build spur roads. He believed that direct appropriations for roads would be more successful than making allowances in timber appraisals, and he did not think that small operators should pay even for the use of federal roads.[53]

Appropriations for timber access roads began to increase, but they were not enough to fulfill Mason's vision of a complete system. Even his estimates soon proved to be too low. He pared his system to a timber-sales policy based mostly on long-term sales ''wherever such larger sales will result in significant substitution of operator built roads and wherever there are no compelling reasons for smaller short-term sales'' and without roads for thinning or salvage operations—''in other words, this minimum estimate would not provide facilities for more intensive

Timber Management than is currently in effect.'' Mason had developed a $75-million road plan for the entire Forest Service. That ran afoul when the ten-year estimates for roads in the Pacific Northwest Region amounted to $133 million.[54]

Mason advised the regional foresters in 1948 that ''schemes to borrow funds to speed up construction of access roads are worth exploration.'' Meanwhile, he advised them to expand the transportation network through a combination of operator construction and requests for direct appropriations. ''Getting timber ready for national emergencies through advance access road construction should be proposed to the newly arising defense planning agencies,'' he suggested. The road program did expand, but not fast enough to suit those who wanted to increase timber sales to the maximum. Road construction defused any lingering feeling that the Forest Service was ''hoarding'' timber, while its limitations were a handy excuse for supply shortfalls. The timber boom had stabilized by 1950, but Regional Forester Perry Thompson in California told a reporter that he was ''determined there will be no logging operations until he gets an adequate sum from Congress to build good roads and maintain them properly.'' Cuttings would be limited to thinning of mature timber in accessible areas.[55]

The ''hoarding'' controversy notwithstanding, the Forest Service was anxious to sell timber. It was because it had always been, since the early efforts to justify the existence of the national forests and to make them self-supporting. The selling boom of the 1920s had made sales statistics the stuff of which job promotions were made. Additionally, the increased demand for timber apparently made an increased sales volume important to the economy and offered an opportunity to defuse critics by showing that the Service met real public needs. Timber sales were the Forest Service's special mission.

There was also the long-standing goal of silviculture—the way to ensure that the forests would produce new wood at maximum capacity was to replace slow-growing old trees with fast-growing young ones. Overlying that consideration was a psychological one. The Forest Service still believed that wood was essential to civilization. In all of the talk about multiple use, timber was in the forefront, especially in the minds of Service leaders who had cut their professional teeth in the timberlands of the Northwest. The foresters believed that other multiple uses should be managed through timber management. Timber restrictions affected wildlife, water, fish, scenery—and, especially, fire control, because cutting and roads broke up fuel masses and afforded entry for fire fighters.

If timber management were to be complete, a significant sales program was essential. The demands of the 1940s finally provided an opportunity to make national-forest sales a significant part of the nation's timber economy. The war had reduced the amount of private timber competing with the national forests and had started a proportional increase in federal harvesting. The postwar construction boom increased the demand further and, despite the negative effects of one last, futile regulation crusade, had created political opportunities for expanded Forest Service programs that supported timber sales. Trucks and other new equipment had eliminated the need for such destructive practices as the old railroad logging. Increased sales could be made safe for the forests and, given a road system, fair to large and small purchasers alike. National-forest timber management was about to come into its own.

TIMBER MANAGEMENT TAKES CONTROL

Before World War II the national forests were mostly custodial institutions, their rangers guarding the resources and perfecting their inventories against the expected time of increased demand. The demand arrived with the war and expanded thereafter. The Forest Service's attention turned increasingly to answering this demand. The agency was greatly decentralized and localistic, but as timber became a larger economic and political subject, inevitably the Washington office attempted to influence what was going on in the field.

Perhaps some intervention was necessary. A new management plan for the Flagstaff working circle on the Coconino National Forest in 1943, for instance, continued the forest's recently developed trend toward countenancing collusion or monopoly for the two companies that were still working in the area. This time the war was the reason. The plan called for heavy cutting in virgin timber, "to remove the volume which cannot be expected to contribute to future costs." Acknowledging the "abnormal demands" of the emergency, the authors said, rather jingoistically, "This war must be won at any cost and the plan must recognize this need and provide means to meet it insofar as it is possible to do so." Within four years, new inventories, increasing stumpage prices, improved operator profits, a growing need to reduce cuts in virgin stands, and operator acceptance of stand-improvement harvests combined to require a revision of the management plan. On Mason's advice, it now called for more intensive utilization of timber that had formerly been wasted and for lighter cutting in virgin timber, on a shorter rotation cycle.[56]

Prescription of the amount of timber that might be cut in a year on each national forest was, until 1945, a power vested in the secretary of agriculture. Watts twice asked the secretary to transfer the authority to the chief of the Forest Service, saying his proposal was "in the interest of reducing paper work and to eliminate a practice that does not improve but tends to slow down the work of the Department and the Forest Service." He got his way on 10 April 1945: authority to approve the annual cutting limitations was transferred to the chief.[57]

If sales were the key to timber management, the Washington office now held the key. Whether it could control lower levels was questionable at the end of the war. To the timber-management staff in Washington, timber did not always seem to receive the attention they believed it deserved at the field level. When the California Region tried to upgrade the status of forest supervisors, removing them from the direct management of the forests to a loftier general administrative position, E. E. Carter objected that "the general idea seems to be to make the Supervisor a little Regional Forester." What bothered him was that the level of professional supervision of resources management was to be reduced under the elevated supervisors, as the region would not raise the grades of timber staff officers. "This seems to me to reflect a lack of appreciation of the importance of the wise handling of the timber resources as the fundamental objective in the establishment of most of the National Forests in R-5," said Carter. "Timber resource management in that Region has been regarded as a side issue to improvements, protection, and recreation. Apparently actual forestry practice in R-5 can be handled by low paid subordinates."[58]

Timber management was the most important matter to Carter, the "fundamental objective" of national forests everywhere; and he was disturbed that anyone might take a different view. In fact, the California national forests had been established for watershed protection, a matter to which the state's people paid a great deal of attention; and even during World War II the forests were overrun with recreationists. But when the Washington office learned that stumpage appraisals in California ran below those in the Pacific Northwest, it sent James W. Girard off on a special investigation, because "this question has long been in the minds of our Division of Timber Management here."[59]

Carter and Mason worked tirelessly to fight what they regarded as attempts in the field to downgrade the importance of timber management or to divert professional staffs to other work. The timber staffs in the Pacific Northwest concentrated on marketing, leaving the administration of sales to lower-level professional sales assistants and subprofessional scalers. Although acknowledging the importance of marketing, Mason wanted the personnel assignments to be reversed, believing that the lower ranks could handle marketing well enough. When the California Region tried to reduce the numbers of district rangers, creating lesser positions for such jobs as tree marking, Carter objected earnestly to the removal of professionals from the woods.[60]

The Division of Timber Management knew where its priorities lay, and in 1945 it proposed a timber budget that would reach an estimated $2.8 million in fiscal 1951, ''with the presumption that the cut of National Forest timber would be stabilized at around four billion feet annually.'' Funding for timber-stand improvement would increase, to complete the work in ten years rather than the twenty that had been proposed earlier. The division proposed that the Forest Service buy and operate its own sawmills, creosoting plants, and other facilities, in order to speed things along, although higher authorities said that such moves would be illegal without the explicit approval of Congress.[61]

Thanks to the Division of Timber Management, the Washington office began to play a larger role in the life of the Regions. When the Northern Region reported unsatisfactory results from spraying in an insect-control project, Mason told the regional forester that ''it is very disappointing to find this same question raised a second year.'' When the Pacific Northwest Region passed the job of marking timber to low-graded personnel, Carter fairly raged that ''unit managership and professional forestry practice are to be misnomers unless the manager controls the marking.'' Rangers should direct it, he maintained. When wartime travel restrictions were relaxed in September 1945, Mason called all regional timber-management officials together for indoctrination.[62]

The leader of the Division of Timber Management was frustrated by the end of 1945; things were not moving as fast as he would have liked. Harvests that year had been three-quarters the volume projected, because of a strike on the West Coast; and on several other aspects of the division's programs in the last year of the war, ''no accomplishments can be reported.'' When it was proposed that funds be transferred from timber to support research, Mason fairly exploded: ''I am strongly in favor of more highly qualified technical personnel to work in the field of timber management. [However,] I am quite critical of the technical accomplishments of the Forest Service, and believe such failures as have occurred are due largely to failure to apply existing knowledge rather than the lack of knowledge. . . . We need to break down the feeling that every technical problem is a research problem to be referred to the Experiment Stations.''[63]

In March 1946, Mason told the Forest Service's personnel chief that there was a ''need for breaking down the feeling that assignments in Timber Management work, or other technical jobs, lead into blind alleys.'' He was equally dismayed when some regional representatives to his timber-management conference had objected to his proposed timber-stand improvement program as being too large. Whatever his frustrations within the Forest Service, Mason more positively perceived a need to sell timber management to the public through aggressive propaganda:

> The Division of Timber Management needs a campaign to sell intensive forestry on the national forests to the people who will be directly affected by our Timber Management programs. These groups are primarily labor, operators, dependent communities, and local consumers. We need to develop an understanding in these groups of the various things which must be done in order to obtain maximum timber yields from the national forests, such as access roads, salvage cutting, area and tree selection, stand improvement, blister rust control, and planting. We need to develop a recognition of the difference in the allowable cuts which can be obtained from intensive forest management. These differentials in yield should then be translated into employment opportunities and maintenance or expansion of community facilities. In a relatively few instances the significance of intensive management can also be tied into local consuming industries, such as the fruit industry of eastern Washington. I believe our initial efforts in this field should be pretty well concentrated in the communities directly dependent on national forest timber.[64]

Planning was also important. The Washington office told the regions in April that "the objective at this time is to have approved management plans within the next 10 years for all working circles in which the timber can be marketed under the economic conditions prevailing during the 10-year period." Mason assigned a new staff member to the "development of some new general instructions on management plan preparation." Nonetheless, and with $17.5 million available for road construction in fiscal 1947 (not counting what the housing agencies would transfer), Chief Watts said he was "convinced that there will be a shortage of forest products to meet national needs for many years to come."[65]

A lot of timber-management plans were revised during the first years after the war, all of which went to Washington for review. A large number increased the allowable-cut levels, often by considerable degrees. The reasons were manifold, including updated information on timber inventory or regeneration, improved harvest methods and equipment, increasing access to timber over greater areas, and, probably, the general administrative sentiment favoring the maximum potential contribution to the national timber economy. Not untypical was the new plan for Coeur d'Alene National Forest, which proposed an allowable cut in white pine that was 300 to 400 percent higher than the levels set in 1939. Plans that did not calculate timber so as to permit the greatest annual allowable cut were returned to the Regions for revision. In rejecting one such plan, Mason suggested that the old shibboleth of community stability ought to be subordinated to the wider objectives of timber management, saying: "It is entirely proper to state in the plan that our sales will be scheduled to provide insofar as practicable for local needs, but we should avoid language which carries an implication that through approval of the plan certain blocks of timber are formally reserved for certain communities or operators."[66]

In July 1946, Mason asked for $3.9 million to prepare inventories and management plans for the 86 percent of the commercial acreage of the national forests that he said had not yet been so studied. The payoff would be increased allowable cuts, higher-quality products, and higher stumpage prices. For the longer term, he asked for a fivefold increase, to $2.6 million, in the annual budget for timber-stand improvement.[67]

The Forest Service's course was well established by 1947. "The Forest Service objective," a Congressional committee was told, "is to grow, sell and have harvested timber crops from the national forests in such a way as will make the greatest contributions to public welfare." The Service said it was willing to do the same on the Indian reservations if Congress wished (Congress did not).[68]

Not everyone in the Forest Service was on the timber bandwagon, however. Mason kept a sharp eye out for digressions. The Southern Region received back a management plan that was disapproved because it lacked "clarity and logic." The Rocky Mountain Region was a bigger worry, because the Colorado timber industry had not shared in the postwar lumber boom. Silvicultural programs, control of logging practices, and road construction on the national forests lagged accordingly. Mason went to see for himself, and he was most displeased to learn that forest supervisors and district rangers spent most of their time working with those who held grazing permits, leaving timber work to inadequately trained or supervised junior staff. He reported that "the Region is preoccupied with range and watershed issues and programs."[69] The facts that the region was in the early stages of a "range war" over grazing regulation and stock reductions and that grazing was its outstanding headache apparently were irrelevant. Nevertheless, Mason had put his finger on an important concern—the quality of work in the woods.

The rapid growth of timber programs had begun to produce its own problems by 1948. In the Northwest, an operator was awarded over $70,000 by the Court of Claims for the destruction of his logging equipment in a mismanaged Forest Service slash-burning operation. That region also had a number of contractors who failed to follow the terms of their contracts. Mason suggested that either the "operation shutdown clause" of the contracts be invoked or bonds be retained until satisfactory performance had been achieved. The Southern Region adopted the practice of dividing its sales into blocks, not releasing a block of timber until conditions were satisfactory in the previously harvested block.[70]

Mason worried about whether enough silvicultural work was being performed in California to prevent cutover areas from reverting to brush. ''We must not let our efforts to build up the cut result in lowering our standards of silviculture or control over cutting,'' he told the Division of Operations. He was also unhappy that financing for ''badly needed'' timber management plans was short. ''Present estimates of costs of inventory and planning work to give adequate coverage of extensive management plans for the national forests total about $3,700,000,'' he said.[71]

Assistant Chief Christopher Granger increasingly came to Mason's support. To Congressman John W. Flannagan, Jr., Granger complained that Congress had cut the Forest Service's timber-sale budget for fiscal 1949, and he told the editor of *Pulp and Paper* that additional road construction into insect-killed timber areas could not be justified until the pulpwood industry had done its part and had justified the action by its demand for the resource. To the regional foresters he said that the Forest Service needed ''more and better timber management plans.''[72] Mason could not have said it better himself.

In the Division of Timber Management, B. H. Payne was dismayed that only a few regions were cutting up to the allowable level, largely because of inadequate roads. L. S. Gross urged the Intermountain Region to promote national forests among small ranchers by emphasizing the employment that the forests offered. In the East, where large timber sales still lay in the future, Gross, reflecting Mason's sentiments, urged the Forest Service to rearrange itself for efficiency.[73]

Locked in controversy with timber men and stockmen, the Forest Service flexed its persuasive muscles to seek support among the public. The old principles of community stability were at least acknowledged in management plans; one for the North Kaibab working circle in Kaibab National Forest stated as its objective the correlation of timber use with the needs of established communities. The 1949 *Yearbook of Agriculture* was given over to the subject of ''Trees,'' and it presented a series of essays explaining various aspects of forest management, emphasizing timber. In 1950, California Regional Forester Perry Thompson said that future sales would be very limited unless funds came forward for roads; he advised the public to write to its Congressmen.[74]

Demand for national-forest timber remained high, so much so that the Ouachita National Forest in Arkansas received bids for old-growth timber at twice the advertised price in 1950. Both demand and silvicultural considerations pointed the Forest Service in the direction of more frequent selection cuttings in ponderosa pine, although variations of systems were legion, according to local circumstances. As the late G. A. Pearson said in his last publication that year, ''Foresters are constantly reminded that since conditions change from one place to another, management cannot be uniform.'' He also said, as David Mason would have, ''Perpetuation of the lumber industry demands large volumes of usable raw wood material in strategic places.''[75]

The Korean War broke out in 1950, bringing with it unknown perils to the forests. According to Lyle Watts, renewed demands generated by mobilization could ''cripple'' Forest Service timber management in the Northwest, where two-thirds of its employees were reservists. ''We are overcutting now,'' he said; ''if war comes we will be dangerously overcutting.'' Someone else told the American Legion at Denver in 1951 that the war *''demands''* careful protection of resources: ''Every stump, stem, and stick has ultimate use in times like these; whole forests, literally, go off to the fighting fronts in time of war.'' The forests came through well enough, however. By 1952 the biggest worry was the accumulation of hazardous, very flammable slash in California forests, where slash disposal had been interrupted while harvests were being expanded during the manpower shortages of wartime.[76]

Timber sales by the Forest Service had expanded enormously by the time Watts retired in 1952. In 1946 they amounted to 2.470 billion board feet. In 1950 they totaled 3.195 billion feet, and two years later, they were at 4.232 billion feet and still climbing. The actual volume of timber harvested on the national forests reached 4.516 billion board feet in 1952.[77]

The pattern of increases each year was firmly established. Timber had always been a large and very important subject in the minds of foresters, particularly those in the Forest Service. But never before had the actual management of timber required so much time and energy and attention, to the point where the timber program itself (rather than its theoretical objectives) seemed to dominate the agency's consciousness and workday alike. Timber was such an active program by 1952 that any ambitious young forester could see that in the Forest Service, timber was where careers were to be made.

CHAPTER FIVE

Adventures in Legislative Sustained Yield

It is recognized that setting up of cooperative sustained yield units involves a certain degree of monopoly. This is unavoidable, and this fact was made clearly evident in the Congressional hearings on this measure. It is my belief, however, that this kind of monopoly, if properly regulated as it is under our cooperative agreements, is of the benevolent type and definitely provides benefits greatly outweighing any disadvantage.

—Lyle F. Watts (1947)

The Sustained-Yield Forest Management Act became law on 29 March 1944. Largely the work of David T. Mason and Edward T. Allen of the Western Forestry and Conservation Association, the act expressed Mason's vision of sustained yield; it was therefore the fruit of industrial support in Congress without much encouragement from the Forest Service. Recognizing that public and private timberlands were intermingled in the West and that a number of timber companies needed new sources of material, the legislation authorized two types of sustained-yield units. ''Cooperative units'' would merge the management of national-forest and private timberlands. ''Federal units'' would reserve national-forest timber in given areas for the exclusive use of local operators. The objective of both was community stability, but at the price of monopoly for the favored companies.[1]

THE FOREST SERVICE AND THE LAW

The Forest Service was of two (or more) minds regarding the implementation of the law. On the one hand, cooperative sustained-yield management offered an opportunity to require conservation on lands that bordered on the national forests. On the other hand, the Service's top leaders were bent on regulation, not cooperation, and they doubted industry's willingness to adopt the Service's principles of sustained yield. Furthermore, cooperative units threatened the Forest Service with loss of control over its own jurisdiction. Last, and by the 1940s less important than formerly, both types of sustained-yield units sanctioned

Lyle F. Watts, chief of the Forest Service during the 1940s and early 1950s. A die-hard proponent of regulation of private timberlands, he found his ego at stake in the sustained-yield-unit program (courtesy Forest History Society).

monopolies and an end to open competition. The Service's first reaction was therefore cautious.[2]

The Simpson Logging Company of Shelton, Washington, applied for a cooperative unit two days after the legislation was passed, and many others followed, especially in the Pacific Northwest Region. The Forest Service, however, discerned more questions than answers. Some officials believed that the industry was reluctant to enter long-term agreements, while others were concerned about the effects of another depression. The general feeling was in favor of management by short-term sales; but on whether operators who practiced good forestry should be favored over others, Chief Watts said "of course" but could not give a "categorical yes." When Watts said that cooperative agreements would favor and perpetuate "responsible private ownership," one regional official ventured that "a number of applicants for cooperative sustained yield units definitely had in mind to eventually turn their holdings over to Government."[3]

At the end of 1944 the Division of Timber Management decided that each Region should develop "at least one tentative sustained yield unit plan for preliminary review. . . . No final plans or formal hearings are anticipated during 1945." By March 1945, seven of the nine Regions had suggested sixty-four possible cooperative units and at least sixty-one federal units. Altogether, the Forest Service had received applications for sixty cooperative units and sixteen federal units.[4]

E. E. Carter thought that it was time to "raise rather sharply the question of whether we mean it when we talk about the stabilization of small communities within or close to a National Forest." Discussing the proposed Big Valley federal unit in California, he pointed out that while it would protect a nearby town, "it would also block other possible purchasers who might wish to reach into this working circle or become established in it, but that is inherent in the basic idea of the Federal unit and the stabilization of the community on the sustained yield basis." In that particular case, he thought that cooperation was impossible with the largest timber owner in the area. Elsewhere, Christopher Granger urged the Southern Region to pursue a possible agreement with the Mansfield Hardwood Lumber Company.[5]

Still the watchword was caution. When a group that included a congressman approached Watts with a proposal for a cooperative unit in the Rogue River area of Oregon, he showed some favor to the proposition but advised that even an acceptable application would require time to prepare for public hearings. He had no intention of holding such hearings while the war was still on.[6]

After reviewing cases in California, E. E. Carter advised the regional forester in San Francisco "to remember that the justification of a case must include a clear showing of public interest due to a danger to an established community that can be given permanence by the proposed action, and that the giving of permanence to that community, and contrasted with the possible use of the timber to support any other or others, is a logical and desirable thing from the viewpoint of public interest."[7] Carter believed that "as the war period passes," the need to respond to applications for cooperative units would become urgent. Meanwhile, Watts objected to a proposed cooperative unit in Oregon because it was in an area where the Forest Service "had contemplated, for the time being at least, continuation of sales on a competitive basis rather than establishment of . . . sustained yield units." He advised the applying industries to continue their plans for expansion without a hard commitment of federal timber. "Efficient mills established at those locations should be in a reasonably good position to obtain an equitable part of the available timber supply."[8]

The Washington office might look askance at the cooperative sustained-yield program, but some of the Regions were downright hostile to the whole idea. Granger told the regional forester in Portland that "some of us here get a rather strong impression that you are distinctly cold on the idea of setting up Federal units" and advised him to temper "what at this distance appears to be an overly negative attitude on the part of the Region."[9]

Not everyone in Washington shared Granger's enthusiasm for the program, however. E. E. Carter stressed "the necessity for our thinking in terms of obligation to the community . . . rather than our obligation to operating concerns." Mostly, he wanted to wait until after the war before considering any such ventures, although he wanted the opportunities for cooperative units to be kept open. His heart obviously was not in the idea, however. He told a regional forester that the objective of the Sustained-Yield Act was community stability; but, "in any event, the act is not one for the relief of cut out lumber manufacturing companies. Probably unnecessarily, I urge that these ideas be kept prominent in the thought of the Forest Service about the application of this law."[10]

Although the war was frequently used as an excuse for delaying the implementation of the Sustained-Yield Act, it was in fact a valid one, given manpower shortages in the Forest Service. Watts felt called upon to make some effort at establishing sustained-yield units, however. He selected one—the application of Simpson Logging Company for lands around Shelton, Washington. That proposal was adopted because it appeared to offer the best possibility of meeting the purposes of the act and because Simpson Logging had "demonstrated a sincere interest in working out arrangements for cooperative sustained yield forest management."[11] A year of negotiations and drafts produced a cooperative agreement and also management and operating plans for federal and private lands. The two parties entered a 100-year contract, planning to harvest 100 million board feet per year until 1956, after which the cut would be reduced to the "allowable sustained yield of the combined ownerships." The objects were better utilization of the forest, higher local employment, less overexpansion, and an end to fluctuations in population and payrolls.[12]

When the proposal was presented to a public hearing in September 1946, it unleashed a storm of controversy. People in the communities that were immediately affected by the proposed unit generally supported the idea; those elsewhere did not. In granting Simpson a monopoly on federal timber, the Forest Service closed the area to others. Farmers, organized labor, small logging operators, and competing companies and communities howled in pain. Small operators would be eliminated, they said; the competitive free market would be destroyed; timber would be diverted from the Puget Sound market; and land uses other than timbering would be restricted. The Grays Harbor communities believed that they had been especially ill-used, while nearly all opponents asked for a reduction in the proposed annual cut. The Grange objected bitterly to the establishment of monopolistic power.[13]

With some minor revisions the Shelton Cooperative Sustained-Yield Unit became active on 12 December 1946. The Grays Harbor region would not be won over, however. When its representatives prevailed upon Senator Warren G. Magnuson to ask for an adjustment in the unit's boundary to leave timber for Grays Harbor, Watts refused to budge, saying: "It is needed to insure sufficient annual cut under sustained yield to stabilize Shelton and McCleary." As for landowners and operators outside the unit's boundaries, Watts said that Simpson Logging Company was free to purchase timber outside the unit "if it so desires."[14]

Christopher M. Granger had been the chief sponsor of the cooperative-unit idea in the Washington office. Once the Shelton Unit was established, he regarded it as a "Christmas present," telling the regional foresters that "now that the ice is broken and as soon as the financial situation clears up we hope that each Region will make real progress in evaluating possibilities for sustained yield units and acting on each pending case."[15]

The hierarchy in the Forest Service maintained that it was dedicated to the stabilization of communities that were dependent upon national forests. Watts put it thus in 1947:

> In establishing the boundaries of national forest working circles, whether or not subject to cooperative sustained yield management, we are emphasizing consideration of community aspects wherever it is possible to do so. We hope to have milling or logging communities so located with respect to the merchantable timber so that the woodworkers will have an opportunity to live at home in permanent communities and to commute to and from work.[16]

That was a lovely picture, one the Forest Service had been painting for several decades. But somehow or other, in actual practice, it seemed that community stability sometimes took a back seat to the Service's timber-management objectives. Opponents of the Shelton Unit had claimed that the stabilization of Shelton's economy threatened their own economy. Watts brushed that off, answering objections by citing the praise of people in the Shelton community. He apparently had decided that the cooperative-unit formula was a way of achieving his objectives for the national forests and private resources alike. When one critic suggested that the program was inconsistent with Forest Service history, Watts replied coldly: "Neither the Department nor the Forest Service takes the position that it has had any unwelcome mandate thrust upon it. . . . The Forest Service promoted the adoption of this measure over a long period and believes that its enactment and its wise application is definitely in the public interest.''[17]

To Watts, support from the immediate community was enough to justify a cooperative unit. Having decided that opponents could be ignored or dismissed as favoring their own interests over the ''public'' interest, he advised regional foresters to enlist local support for cooperative units. It was not to be, however. The Forest Service attempted to form additional units at Libby, Montana, and Quincy, California; but it quickly ran into a storm of opposition from small operators, organized labor, and communities that would be adversely affected. Overriding everything were objections, from self-interest and on legal and ethical principles, to the monopolistic nature of cooperative units—the favoring of one to the exclusion of all others. The Shelton Unit was the only cooperative program established under the 1944 legislation. Local opposition scuttled even the Department of the Interior's attempt at cooperative sustained yield in Oregon.[18]

Frustrated in its attempts to establish more cooperative units, the Forest Service turned to establishing units that were more within its power—the federal sustained-yield units involving only national-forest land.[19] Federal units were not to come easily, either, because they involved intervention in local economies that troubled some people in the Forest Service. Worse, in many minds, were the possible enemies who could be generated for the whole sustained-yield-unit program. In reviewing a plan for the Sitgreaves National Forest in Arizona, even Christopher M. Granger was bothered: ''In addition to considering whether such an obligation would be worth the additional community benefits, we should also consider whether the additional benefits to Heber would more than offset the disruption of established community values at Safford, where the Company's finishing facilities are now located. If Safford is a better place to live than Heber, should we force the Company to move its employees now residing at Safford?''[20]

The line between the promotion of community stability and the high-handed manipulation of private affairs was indeed a thin one, and that made development of sustained-yield units more difficult. The Forest Service managed to create only five federal sustained-yield units, each of which became a perpetual source of frustration and complaint, reserving 1.7 million acres of national forest land in Arizona, California, New Mexico, Oregon, and Washington.[21]

The Grays Harbor Federal Sustained Yield Unit in Washington reflected the Service's retreat. Stung by cries of outrage from the Grays Harbor area when the Shelton Unit was proposed, the Forest Service first sought a similar arrangement around Grays Harbor. Despite the good offices of William B. Greeley, the Forest Service could not come to terms with local industries. Ultimately, the best the Service could offer was a wholly federal unit, designed to supply local needs and to encourage both reforestation and product remanufactures. The unit became active in November 1949. Nearly a year later the Lakeview Federal Sustained Yield Unit in Oregon came into being under similar circumstances.[22] Another federal unit at Big Valley, California, endured for years on paper but never operated successfully.[23]

SUSTAINED YIELD AS SOCIAL ENGINEERING

The 65,000-acre Vallecitos Sustained-Yield Unit on the Carson National Forest of Northern New Mexico began as a social-engineering project with some peculiarly diverse purposes—range improvement, community stabilization, and timber sales. The cutting level was low on the forest after World War II, and the Region wanted the level raised, starting in 1947. The Forest Service came up with a plan to support the development of a sawmill and box-and-stock factory near the town of Vallecitos. According to the plan, this would draw upon the timber and "raise the economic well-being of these small farm-stock owners." The main purpose of the plan, however, was to provide compensating employment income to subsistence-level graziers whose federal grazing allotments were about to be reduced. Sustained-yield forestry, in other words, masked a plan for range improvement.[24]

In assuming charge of the national forests of northern New Mexico, the Forest Service intruded upon a society of medieval peasantry (legal peonage had existed in New Mexico as recently as the 1860s). Very poor people scratched subsistence from an ungenerous landscape, much as they had for centuries. Foresters, who were interested in restoring overused rangeland, inevitably threatened the very survival of the people in the region. The Forest Service was not insensitive to the problems that it would create, but it was determined to have its way regardless.[25]

When the Forest Service reduced grazing allotments in 1947, Pedro Martinez told the regional forester that the agency did not have the interests of the "poor people" at heart. Nor did he have any confidence in the proposed sustained-yield unit. He predicted that "strangers" would come from other states. They would say: "The hell with the poor people of Vallecitos, Petaca, and Cañon Plaza. We got the money and we are going to drive them out." He predicted the end of his people's way of life, because "you have taken away from us the rights of our predecessors who were permitted 25 head on a free permit and 10 head of livestock, general, on a free permit and 60 free sticks of wood for our own use. And where are they? Dead."[26]

It remained to be seen whether technicians who were oriented toward timber and grass could manage people adeptly, when the people were from an ancient and nontechnical culture. It also remained to be seen whether a federal sustained-yield unit would be an effective tool for the management of people (or community stabilization, as it was called). Affairs did not start smoothly. Shortly before the unit was to be activated, the firm that was to develop the sawmill and stock plant pulled out, and the Forest Service started a search for a replacement. On 21 January 1948, Chief Watts established the Vallecitos Federal Sustained Yield Unit; on that same day, he amended its policy to include the communities of Petaca and Cañon Plaza, along with Vallecitos. The policy provided that local people could purchase the locally produced lumber, "but we have no basis for requiring that such sales be made at any specified price."[27]

Vallecitos was the first federal sustained-yield unit established under the 1944 legislation; it was also the smallest, with an average annual cut of 1.5 million feet. The management plan called for selling the unit's timber to maintain steady employment for local resident labor and to provide lumber for local requirements. The Vallecitos Lumber Company was established and designated as the "approved responsible operator"; it was charged with installing, maintaining, and operating a "primary manufacturing plant including planer" at or within a mile of the village of Vallecitos. Ninety percent of the employees were to live within ten miles of the plant. The company's designation soon was terminated because of noncooperation, and in 1952 the designation went to Jackson Lumber Company of Vallecitos.[28]

Operations in the unit suffered from labor/management difficulties, which arose from cultural conflicts, divergent priorities, and poor relations between the designated operator and the Forest Service. Only five men showed up to operate the mill in the fall of 1952, and the timber staff officer for Carson National Forest investigated. Two storekeepers told him that "the fault was probably with the employees not wishing to work in cold weather, wanting to cut wood for their homes, and possibly wishing to be released to go on relief." Another storekeeper added that the crews believed that they were not being paid properly for the volume they harvested. In addition, the employees preferred to cut only in pine, while the company wanted mixed conifer production. A year later, however, the regional office declared that "the advantages to the community are clearly evident."[29]

Relations between the operator and the Forest Service deteriorated. In 1955, Jackson Lumber Company engaged a former Forest Service man as "consulting forester," to work with the Forest Service in "planning and supervising the woods operation." He soon began to complain about "the feeling of continual antagonism and bickering which prevails between the Forest Service and Jackson Lumber Company at Vallecitos." He accused Service personnel of finagling on the scaling of logs; of "inciting unrest or dissatisfaction among the workmen"; of meeting with the people of Vallecitos "to stir up dissent against the Company"; and of arbitrarily threatening to shut down the operations. The Forest Service dismissed his complaints as being merely the spite of a disgruntled former employee.[30]

The company also failed to meet its quota of 90 percent local employment; about half the labor was imported, and local people complained. The company, however, was unable to meet the quota. The Forest Service propounded new guidelines, extending the labor pool to eight communities and to "adjacent rural areas," in the process making residents of El Rito ineligible.[31] Jackson Lumber appealed, and the Department of Agriculture held public hearings. Testimony revealed that the community of Ojo Caliente, many of whose residents worked for Jackson, was divided by the 10-mile labor-pool boundary. Furthermore, Jackson and its employees fell out over piecework versus hourly payment for work in nonfederal timber, and the workers tried to establish a union to demand an hourly wage. The company fired some of them and then brought in workers from Texas.[32]

The main point of dispute was the question of whether Ojo Caliente was a community that ought to be included within the labor-pool boundary; if it were, the company would have no difficulty in reaching the level of 90-percent local employment. Ojo Caliente was an old colonial land grant that took modern form as a succession of small holdings that extended ten miles down a valley. To the Anglo foresters that was not a community; it was a rural area. The company and the Forest Service fell into a series of arguments and appeals over the drawing of the 10-mile boundary. Finally, in 1956 the Forest Service agreed to "consider persons residing within the Ojo Caliente Land Grant to qualify as local laborers for purposes of complying with the terms of the two non-competitive timber sale contracts."[33]

That did not end matters. The sustained-yield exercise in people management was a futility, so the Forest Service held hearings on whether to discontinue the unit. Residents of six villages asked that if the unit were to be continued, the policy of having 90 percent local employment (that is, exclusion of Ojo Caliente) be enforced.[34] The hearing became so hostile that Jackson Lumber Company did not make a presentation, contenting itself with a statement for the record, which it submitted later. Early in January 1957 the chief of the Forest Service noted that the people generally favored the continuation of the sustained-yield unit, but not with Jackson Lumber Company. So he appointed an advisory board to offer recommendations on labor questions.[35]

The Carpenters and Joiners Union bombarded the Forest Service with complaints that Jackson Lumber was importing workers from elsewhere. Senator Dennis Chavez tried to work out a compromise formula, but the union claimed that the company did not even conform to that. When the Forest Service developed a clause for timber-sale contracts, requiring local employment on its terms, the company demurred unless it could determine whether sufficient local labor was "available." On 2 May 1957 a Forest Service delegation went to Vallecitos to tell Jackson that the clause would stand. Jackson "said that if it had to be that way he was through—that he could not and would not sign a contract with that clause in it." Jackson's mill burned to the ground that night. Three days later the company refused to work under the labor restrictions, and on 23 May the Forest Service revoked Jackson Lumber's designation for the sustained-yield unit.[36]

The Vallecitos Federal Sustained-Yield Unit managed to stay on the books but had little effect on the ground. Another operator was eventually located, but its mill burned down in 1963. In 1966 an inspection report said that the unit had "failed conspicuously to meet its objectives" and recommended that it be terminated, with a caution about "the delicate public relations situation involved." The question became critical the next year, when a fire on the Carson National Forest left a lot of timber to be salvaged. The timber had to be put on open sale because the unit was not functioning.[37]

The Washington office wondered "what to do about an inoperative unit which is obstructing use of National Forest land for benefit of local communities." It was known that the people of Vallecitos would not take the termination of the unit lightly. By 1969, another local operator, Duke City Lumber Company, was pleading for a modification or termination of the unit to open its timber to the firm.[38] Duke City Lumber had a mill that was operating in Vallecitos, and the company had been buying national-forest timber in regular sales since 1962. It needed access to the unit's timber, but the people of Vallecitos would not budge. The answer was to bring the two sides together. Duke City Lumber applied for designation as the "approved responsible operator"; the workers in Vallecitos consented at a public meeting; and on 4 April 1972, Duke City Lumber became the operator of the sustained-yield unit. Its sawmill burned down in 1977.[39]

From every standpoint the Vallecitos Unit was a dismal failure. By the time the Forest Service had acknowledged the mistake, it could not be corrected; local people simply would not go along with termination, for whatever reason. As an exercise in sustained-yield timber management, the unit that was dedicated to idle or inflammable mills actually inhibited the systematic management of the timber on the national forest.

Much of the error lay in the unit's vague and conflicting purposes. The national forest's immediate concern was the reduction of livestock grazing by people who could not afford to give up a calf. The forest staff also had its own internal problem—unhappiness because its timber sales were at a low level. Superficially, the movement for sustained-yield units appeared to answer both concerns. It did not, however, because sustained-yield management turned less on matters of economics or forestry than on the sentiments of people. People can be cantankerous, and they certainly were so at Vallecitos. Even the ageless magic of "sustained yield" proved wanting in the real world. Slogans and labels could not hide the fact that national forests would ultimately be managed in the way that the public wanted them to be managed, not in the way that the foresters thought they should be. The conceivers of the sustained-yield unit walked into northern New Mexico as if with their eyes shut. Their unit never stood a chance.

SUSTAINED YIELD AS ECONOMIC BENEVOLENCE

The last federal unit established under the 1944 legislation—the Flagstaff Unit, created in 1949 on the Coconino National Forest in Arizona—came to a better end. It was terminated.

In 1939 the Forest Service estimated that half the population of Flagstaff was dependent upon the two sawmills drawing on the Flagstaff Working Circle. As early as 1943, E. E. Carter suggested that that region might be a good candidate for a sustained-yield unit. By 1946 an application had been received and a study was under way. A unit was recommended the following year, because the joint operation and market splitting of the two local firms—collusion that the Forest Service had promoted for several years—threatened to come apart, with possibly adverse effects on the local economy. Forest Service officials thought a sustained-yield unit would be a means of stabilizing the community and regulating industrial harvesting. Mostly, it appears that Acting Forest Supervisor Roland Rotty was thoroughly enamored of the idea of sustained-yield units. The regional forester, however, regarded the whole thing as an unwarranted gift to two local industries.[40]

The national forest wished to stabilize the two firms by letting 100-year contracts that would give them a monopoly on the timber of the Coconino National Forest. It hoped to use its management plan to divert them from railroad logging to highway transport, and above all it wished to avoid the introduction of seasonal communities by smaller operators who might outbid the established firms. The regional office turned down the 100-year contract, and the Washington office decided that Flagstaff did not qualify under the Sustained-Yield Act. By 1947 the applicants had enlisted political assistance, and the Forest Service was discovering that the true degree of the community's dependence upon timber could not easily be measured. While the local officials kept promoting the idea, the Washington office grew increasingly sensitive to the monopolistic aspects of the plan.[41]

Suddenly, in 1948 the Washington office turned around and told the Region to go ahead with public hearings on a sustained-yield unit, offering suggestions on how to load the hearings with favorable witnesses. Meanwhile, with the Forest Service acting as arbitrator, the two large concerns formally divided the unit's timber approximately half and half, while the Service devised a 20-year plan that would serve the two large concerns and one small door-and-sash mill (the owner of which was soon eager to sell out to the large combine for a good price). The plan would give the big industries a nearly permanent resource that they could exploit over the counter in a number of imaginative ways. Roland Rotty was not worried: "Once I was much concerned about this, because I did not see why any private citizen should profit by dealing in something that belongs to all the people. Since then I have come to realize that this is the case throughout the entire business world. . . . I see no reason to get excited about this. We should go ahead and conduct our timber management business without trying to prevent somebody making a profit by selling out."[42]

The Washington Office was very concerned about possible opposition to the proposed unit. Christopher Granger advised the regional office to "anticipate" opposition "and be prepared so far as possible to have them counteracted at the hearing." He did not want another fiasco like the recent outcry that had killed a proposed unit in California. "Our position is that it is the Forest Service responsibility to make sure that everyone has a good understanding of the proposal but not to engage in public debate at the meeting." The solution was to line up supporting witnesses.[43]

Public sentiment for open competition threatened the Service's plans for Flagstaff. The current timber plan called for competitive bidding, and the regional office feared that high bidders might take timber to mills located elsewhere. Establishment of the sustained-yield unit would prevent that. In the interim, the Region promised its industrial clients in Flagstaff that it would "continue to supply stumpage to established plants dependent on national-forest timber much as those at Flagstaff. This can be done under war power acts, as long as prospective purchasers bid the OPA [Office of Price Administration] ceiling prices."[44]

During the more than a year that the Flagstaff sustained-yield unit had been under discussion, any misgivings in the Forest Service were swept away by high-level enthusiasm for the whole program for sustained-yield units. On 25 October 1948, Chief Watts approved the proposed unit. His only real concern was opposition, which he told the regional forester to defuse: "When you are convinced that the proposal will be actively supported locally, you are authorized to proceed" with public hearings.[45] Watts feared the public's "strong sentiment for making all Federal Units competitive." He underscored his belief that the maintenance of year-round operations by the two mills at Flagstaff would not be possible if they had to face competitors. He wanted it clearly explained, however, that 15 percent of the allowable annual cut would be set aside for competitive bidding by other purchasers.[46]

Charges that the Forest Service was promoting monopolies clearly worried Watts, and he grasped for responses: "One of the strongest arguments is that the two large sawmills offer an opportunity for local laborers to have a choice of employer. Flagstaff is not at all a one company town." Considering that the two mills operated substantially as a combine and that the Forest Service's plan required them to do so, that was a remarkable statement. Watts believed that "community stability" depended upon "maintaining the equivalent of the manufacturing facilities now in operation"—which meant no more as well as no less.[47]

The Region did its spadework: it prepared an elaborate booklet justifying the sustained-yield unit, sent officials out to make speeches, and obtained favorable comment in the press. But it did not win everyone over, particularly when one of its supporters praised the Forest Service for trying to eliminate the small mills that "nibble at the flanks of the pine forest." The Western Forest Industries Association, a group of small operators who were opposed to the unit, said: "Fundamentally our association believes that government agencies should be concerned only with the stability of wood using communities and not with that of individual operators." Nine small operators banded together as the Coconino Small Mills Association, thus belying the assumption that the timber industry of Flagstaff was represented by three firms, two large and one small. The nine bought a full-page advertisement in the Flagstaff newspaper, under the headline "Sustained Yield or Sustained Grab?" The stage was set for the public hearing.[48]

The meeting took place in Flagstaff on 2 February 1949. A parade of civic leaders, bankers, organized labor, and others came forward to promote the sustained-yield unit. Their chief argument was that stabilizing the two large mills by guaranteeing their timber supply and protecting them from competition would attract development capital and would permit them to grow. Small operators, understandably, bemoaned the anticompetitive nature of the unit and predicted the death of their own industry. Afterwards, the regional forester suggested that the amount of the annual set-aside for competitive sale be increased, but Watts turned this idea down. Small operators were told that they would receive somewhat more timber than they had taken in recent years and that they should console themselves with the thought that the two big Flagstaff concerns would not be allowed to bid against them. When one of the two giants tried to buy out a small operator, the Forest Service stopped the action as being a violation of the unit plan.[49]

The Flagstaff Federal Sustained Yield Unit set aside a perpetual timber supply, free of competition from other firms, for Southwest Lumber Mills, Inc., and Saginaw-Manistee Lumber Company, working together and dividing the resource between them. The arrangement worked well enough, in the Forest Service's opinion, as the Region reported in 1953:

> The establishment of the Flagstaff Unit has permitted the community to maintain itself so far as the lumber industry is concerned on a comparatively even keel. Whether this would have been possible without the Unit is open to considerable question since the timber now being sold is competitively available to plants at other locations. We cannot claim, therefore, that conditions have been improved as a result of the Federal Unit program, but they have not deteriorated as they might have without it.[50]

Watts had answered charges of monopoly by saying that the two colluding operations were preferable to having only one. That went out the window in 1954, when Southwest Lumber Mills bought out Saginaw-Manistee, and the Forest Service made the new giant the sole "approved responsible operator" for the Flagstaff unit. The company embarked on an expansion program. In 1956 an "approved responsible operator" was designated for pulpwood, a designation bought and sold repeatedly thereafter.[51]

Sentiment soon emerged nationally for the repeal of the Sustained-Yield Forest Management Act, which was increasingly regarded as socialistic, and bills were introduced in Congress. The Forest Service repeatedly had to review the value of the few existing units. Regarding Flagstaff, the Service reported in 1956: "It is impossible to demonstrate that the establishment of the unit has had any effect upon community developments or expansion. Improvements have occurred but no basis exists for attributing them to the existence of the Federal Unit.'' That seemed to be no reason to discontinue it, however. The next year it was reported that the number of small operators in the area had dwindled to four.[52]

Southwest Lumber Mills was certainly doing well. In 1957 it received a 30-year pulpwood contract, involving about 6 million cords, to be harvested beginning in 1962. By 1960 the company had attracted $40 million in capital to finance its expanded ventures, the financial commitments being contingent upon the continued availability of the sustained-yield unit. Meanwhile, the Forest Service tried to attract a newsprint plant to Flagstaff to support ''community stability'' further.[53]

In 1962 the Forest Service granted to its designated operator—the burgeoning enterprise that by then had been renamed Southwest Forest Industries, Inc.—the authority to bid on all competitive sales that were offered inside or outside the Flagstaff Federal Sustained-Yield Unit, until further notice. That ended the reservation of 15 percent of the annual cut for smaller competitors, who thereafter faced direct competition from a very profitable giant, which could now gobble them up.[54]

Southwest Forest Industries bought out the Kaibab Lumber Company in 1965, and it received permission to transfer Kaibab's purchases of national-forest timber to itself.[55] Southwest clearly enjoyed significant advantages from its monopoly on the sustained-yield unit's future timber; it became the dominant forest industry in a region that extended beyond the Flagstaff unit (the corporate headquarters, in fact, were in Phoenix). Relations with the Forest Service were exceedingly close. When the national forest revised its timber-management plan in 1965, it met with corporation officials to map out a plan of action for breaking the news to the public. The announcement of Southwest's purchase of Kaibab Lumber, the Forest Service advised, should explain that it had happened ''in order to bring local mill capacity in line with available timber.''[56]

The original purpose of the sustained-yield unit had purportedly been to ''stabilize'' the community of Flagstaff. The adopted course of action had involved the questionable selection of a combine of two firms (which had soon merged) for preferential treatment; an element of distaste for small operations also figured in. Southwest Forest Industries extended its monopoly and built a veritable economic empire. By 1968 it was very jealous of its prerogatives.

That year, the C. T. Bunger Lumber Company reconstructed an old sawmill, and Passalacqua Lumber Company erected a new facility. When they petitioned for Southwest to be excluded from the 15 percent of annual cut formerly reserved for small operators, the regional forester agreed. Southwest hit the roof. It claimed that it had ''paid dearly'' for

Kaibab Lumber in order to institute double workshifts at its main plant. It was also building a large particle-board plant at Flagstaff to increase timber utilization, and it claimed that the loss of the 15 percent of the sustained-yield unit's annual cut would be "economically disastrous." To Southwest, its new competitors were "a haphazard and inefficient operation capable of cutting only three to four million feet per year."

The regional forester denied the appeal. Allowable cuts might rise or fall in future years, he said. As he wrote to the firm's vice-president: "For this reason we have not encouraged industry to build beyond the capacity of the Unit to sustain. In past years members of your company have discussed overcutting on the Unit to permit two full-time shifts in the mill with the realization that after a few years the allowable cut would drop to a one shift basis." Regional Forester William D. Hurst apparently believed that the Forest Service had created a monster that it could no longer control. He hinted that a complete review of the sustained-yield unit could be called for if the company were to insist—a veiled threat that Flagstaff might no longer qualify under the original legislation.[57]

The extent to which forest officers had been willing to assist the designated operator in the name of community stabilization came to light in 1969. At Flagstaff, the Forest Service had consistently adjusted destination calculations in timber appraisals to produce lower appraisals in the sustained-yield unit than those prevailing in competitive sales. The Washington office said, "It is our strong belief that there is nothing in the Sustained Yield Forest Management Act which either requires or permits timber in Sustained Yield Units, either Federal or Cooperative, to be appraised any differently than if it were not in a unit." Furthermore, it was a "false premise" that community stability was to be promoted "by assuring a supply of timber to dependent communities but by a price concession as well. This . . . simply can't be read into the Act." Competing timber industries had lodged justifiable complaints about favored appraisals in the sustained-yield unit. The regional office conceded that "the effect of appraising this timber to Flagstaff is to give Southwest . . . a price concession as well as providing protection against competition. . . . Our job is to establish fair market value, not to sell at reduced prices." Advising Washington on corrective measures, the regional staff said: "We will have to . . . admit that we drifted into our present position without considering all of the complications." The Coconino National Forest was ordered to correct its appraisal concessions in the sustained-yield unit; nevertheless, accusations of unfair pricing continued.[58]

The national forest also had to reevaluate its policy in regard to sales to small operators in the 15 percent set aside for them. The regional office declared that the intent of the set-aside percentage ''was to afford protection of the small operators who were operating on the Coconino in 1949.'' However, whereas there had been eleven such operators in 1949, by 1969 all had ''passed out of the picture with one exception,'' because the chief had refused to support small operators by allocating timber. ''The whole intent of the Act was to stabilize communities and not individual operators except as a device contributing to community stabilization,'' Watts said in 1949. With Southwest's having driven out the small operators and with the giant now resisting attempts by new small firms to reclaim their 15 percent, it seemed questionable whether the intent of the legislation had been observed at Flagstaff. In any event, the Region decided to advertise sales on the 15 percent without reference to the size of the purchaser.[59]

The Forest Service made a periodic restudy of the Flagstaff Unit in 1970 and concluded that ''it is difficult to assess what competition would be if the Unit did not exist.'' Flagstaff's economy was only about one-fifth dependent on forest-products industries. Nonetheless, the report recommended continuing the unit, because it ''has been successful in fulfilling all of its objectives and the purpose for which it exists.'' F. Leroy Bond in the regional office disagreed, however, because the timber industry was no longer the ''key to economic stability'' in Flagstaff.[60]

Southwest Forest Industries had only three small competitors working the 15 percent set-aside area in 1970. The unit continued to exist for several more years, with periodic revisions of the management plan and amid growing public complaints. The unit came under question again in 1977. Predictably, Southwest promised ''disaster'' and ''disruption'' if the unit were to be discontinued. J. D. Porter, president of the Western Pine Industries Association, had another opinion:

> We oppose the continuing of the sustained yield unit on the grounds that it favors big business and puts the sustained yield operation at a definite economic advantage over the ''outside the unit'' operators. This economic, non-competitive advantage gives them unfair advantages at other sales on other forests where they, then, compete at open biddings, having the negotiated or non-bidding value of timber to support their bidding advantage on other forests.[61]

Late in 1977, Forest Supervisor Michael A. Kerrick of the Coconino directed that a ''white paper'' on the Flagstaff unit be prepared. ''It might be that eliminating the designated operator would serve to promote open and fair competition,'' said the paper, which also suggested that alternatives should be submitted to the public. In transmitting the paper to Washington early in 1978, the regional office advised that at least one firm was thinking about challenging the Flagstaff unit in court; and it predicted that Western Forest Industries Association would join any such suit. "There is a possibility the Unit could be challenged successfully," said the Regional Forester, "for not meeting the intent of the law; specifically that portion that speaks to following the usual procedures in selling timber."[62]

Opponents went to work, and the Forest Service began to hear from members of Congress in 1978. The Service initiated another review, which concluded that "the Unit discriminates against other communities, businesses, and citizens of adjacent communities" and that it ought to be abolished. Public response to the review included such phrases as "Enough is enough" and "Fidelity to SFI [Southwest Forest Industries] is like 'DCS' (Damned Chrysler Syndrome)," the latter a reference to the contemporary public bailout of the failing Chrysler Corporation. The unit's days were numbered.[63]

Public hearings on the future of the Flagstaff Unit were held in January 1980. Southwest Forest Industries and its friends tried their best to guard its privileged position.[64] It was to no avail, however, for the nation would no longer tolerate unseemly governmental interest in the care and feeding of a thoroughly prosperous firm, especially when federal action worked to the disadvantage of others. In May 1980, Forest Service Chief R. Max Peterson said to the regional office: "We concur with your recommendation that the Unit be dissolved." The regional office's press release said that opinion at the public hearing had been about "equally divided," but with Flagstaff no longer being "primarily dependent on the sale of national forest timber," the continued maintenance of the monopolistic unit could not be justified.[65]

A revised policy statement for the Coconino National Forest brought it back into the mainstream of the Forest Service in regard to timber sales, by emphasizing advertised, competitive bidding. Southwest Forest Industries' current contracts continued, but timber produced under them could now be taken away from Flagstaff for primary manufacture. Eighty-five percent of the forest's timber would be offered without preferential bidding under the rules of the Small Business Administration, but it had to be manufactured in the Flagstaff region. The remainder of the timber would go in preferential bidding by small business. Southwest Forest Industries' attempt to reverse the action that terminated the unit was tossed out of court.[66]

UNLEARNED LESSONS

Dubious in its justification and questionable in its history, the Flagstaff Federal Sustained Yield Unit became a sore embarrassment to the Forest

Service before it was discontinued. It should never have come into existence. Even the Service's own studies during the late 1940s had questioned whether Flagstaff's survival depended upon a sustained-yield unit, let alone the peculiar construction that actually emerged.

The Coconino National Forest had a history of encouraging dominant timber firms at the expense of competition, so the arrangement that served Southwest Forest Industries' predecessors was not utterly foreign to the area. Local forest officers had a longstanding interest in stabilizing both the economy and the practices of the local timber industry; they perceived cooperative large operators as being more conducive to that end than small or seasonal operations. During the 1920s and 1930s the forest was divided into exclusive territories for major purchasers. The sustained-yield unit, handing 85 percent of the entire forest to a single combine, was a logical continuation of previous developments.

The Flagstaff Unit, however, was definitely not the kind of arrangement that the framers of the 1944 legislation had foreseen. They had wanted to support communities by stabilizing the flow of natural resources to dependent industries. The Flagstaff program attempted to determine the economic structure of the community. Worse, it assumed that an industrial monopoly was the secret to Flagstaff's stability; in effect, it supported a corporation instead of the community. The Forest Service's traditional fear of monopolies was, for the moment, forgotten.

The Flagstaff Unit did not come into being solely from local considerations, however. Certainly the local staff of the Forest Service hopped onto the sustained-yield bandwagon to achieve what they had failed to do when they had proposed 100-year sales to the combine (the unit by another name). Nevertheless, the real creators of the unit were the Forest Service's leaders in Washington. Christopher M. Granger was a great booster of sustained-yield units, but they were his special charge. It was Lyle Watts, more than anyone else, who saw the Flagstaff Unit into existence—questionable origins, conceptual warts, and all.

Watts's (as well as the Forest Service's) ego may have been at stake. He had first greeted the sustained-yield program with caution, especially while the war continued. Once he had taken the measure of the law, however, he became an enthusiastic supporter. Sustained-yield units appeared to serve a number of cherished Forest Service objectives. Properly framed, they could promote community stability, justify the existence of the national forests, and force operators to harvest conservatively on private lands. Watts was able to achieve only one cooperative unit, however. The Shelton plan sparked outrage from other "communities" that were disadvantaged by the arrangement. More effective opposition from a cross section of the public scuttled all other attempts at cooperative units. Facing unaccustomed defeat, Watts and the Forest Service turned their attention to federal units. Those also ran into opposition, and ultimately the Regions quit trying.

Flagstaff was different. There, organized opposition seemed to be minimal, while potential support was widespread. The proposed unit was a dubious prospect, but its establishment would be a vindication, a saving of face, for Watt and his objectives. Guaranteeing that the Region had greased the public-relations skids, the chief told the Region to launch its proposal. When the Flagstaff Federal Sustained Yield Unit came into being, the entire program was somehow justified.

Watts was a realist, so he quit urging that more units be established. He could not bring himself to admit that the program was dead, however. In 1950 he told a newspaper reporter that there had been ''some difficulty with sustained yield units but my mind has not changed one iota. That was one of the finest pieces of legislation passed.'' He did admit that no new units were contemplated immediately. In 1953, after Watts had retired, the Department of Agriculture effectively buried the program by requiring communities, rather than the Forest Service, to initiate proposals, reserving approval to the secretary of agriculture. Four years later, the secretary announced that it was departmental policy to establish no new sustained-yield units of either type, though established units would be ''continued for the present.''[67]

Frustrated as it was, the program of sustained-yield units was the Forest Service's grandest and most systematic attempt to promote community stability through the management of national-forest resources. The program was doomed, however, by faulty conception, compromise with monopoly, human nature, and questionable intervention into local and private affairs.

The Forest Service embarked on the program while it was locked in its last feud with industry over the regulation of harvesting on private lands. To the extent that cooperative units might be used to control private operations, the program fit in with the general scheme. Unexpectedly, the units ran into strong, often emotion-laden opposition—and not just from the timber industry. Conservation groups, organized labor, civic organizations, and other traditional supporters of the Forest Service rose up in outrage and defeated the proposals one by one. There were doubtless some bruised feelings among those who were leading the campaign for establishing units, as they heard themselves castigated as enemies of the little fellow, disrupters of communities, friends of monopoly, and altogether too cozy with certain industries. The Forest Service had never heard such things from its friends before. Its political enemies previously had been well defined and rather easy to dismiss as self-serving—primarily large industry, especially during the Progressive and New Deal eras, when large industry was a national whipping boy.

Unexpected opposition from unexpected sources should have given the agency pause, causing it to reexamine its objectives and to redefine its vision of the public interest; but it did not. In the debate over regulation, the agency, sure of the rightness of its cause, brushed aside any objections from its opponents. The Forest Service did the same on the matter of sustained-yield units. It failed to consider whether the objections were valid; and it did not really even listen to what its opponents had to say.

By not distinguishing the opponents of regulation from those of sustained-yield units, the Forest Service did itself a disservice, for a time was approaching when many public interests would voice differing opinions about the Forest Service and what it was doing. Criticism of government would be the rule of the day. The wreck of the program for sustained-yield units might have been good preparation for what lay ahead. Instead, the Forest Service looked inward, content to believe in the purity of its ideas and in the wrongness of those who disagreed with them. The world was changing again, and again the Forest Service, heeding its technocratic muse, was slow to notice.

CHAPTER SIX

Multiple Use, Sustained Yield, and the Winds of Change

A timber sale—a timber-harvesting program on the national forest—is no longer a matter just between [the logging industry] and the Forest Service. . . . More people are seeing first hand the results of our management. More are ready to criticize if they do not like what they see.

—Richard E. McArdle (1958)

The decade of the 1950s presented the Forest Service with something it had not faced since 1933—a Republican in the White House who was averse to regulatory crusades founded on the alleged moral inferiority of American businessmen to federal bureaucrats. As a result, the foresters' energies were diverted into a campaign for "multiple use," an old idea that was given new dressing and brought to the fore. The Service's focus on timber merged conveniently into the new watchword, and by the end of the decade the Forest Service had a more modern vision of its mission—which provided a new way of protecting the agency against criticism from any position. But in all its talk about multiple use, the Service failed to appreciate fully the rise of multiple demands for national-forest resources, timber and otherwise.

THE LAST STRUGGLE FOR REGULATION

The Cooperative Forest Management Act of 1950 authorized the secretary of agriculture to cooperate with state foresters in assisting private owners of timberlands. That was a small reward for Watts's several years of bitter struggle for regulation. So he kept up the campaign, dispatching Assistant Chief Edward C. Crafts to Yale University in December 1951 to deliver a speech on the need for regulation. Crafts predictably said that the regulation of private timber was essential to the public interest and that the alternative was increased federal ownership of timberlands.[1]

Edward P. Cliff recalled that "this incident created a stir in the timber industry."[2] That was an understatement. Industry reacted with outrage, and the issue became involved in presidential politics. The 1952

Edward P. Cliff as assistant chief during the 1950s. The Service's chief spokesman for sustained yield and multiple use, he later presided as chief during the eruption of the clearcutting controversies of the 1960s (courtesy Forest History Society).

platform of the Republican party called for the "restoration of the traditional Republican public land policy," while that of the Democratic party weaseled about the "wise use" of natural resources. Candidate Dwight D. Eisenhower flatly opposed the "federal domination of the people through federal domination of their natural resources," while his chief adviser, Governor Sherman Adams of New Hampshire, was a timberman who opposed regulation.[3]

The leadership of the Forest Service changed hands at the height of the regulation furor. Watts retired on 1 July 1952, leaving behind this advice: "Never forget the basic philosophy on which the Forest Service has grown great, 'The greatest good of the greatest number in the long run.' The greatest number of people are little people, and they are the ones who need to be remembered."[4] Watts's successor was Richard E. McArdle, a veteran of a widely varied career in the Forest Service, lately assistant chief in charge of State and Private Forestry. Leaders of the forest-products industries looked at McArdle's ascension with some caution. A few feared that, as Watts's hand-picked successor, McArdle would continue the crusade. Others were more hopeful. One of them said: "Here is our opinion of McArdle—he is a 'smart' boy and unlike Watts will shape his course to fit in with any administration. I mean of course, his principles are flexible."[5]

McArdle may not have been so flexible in principle as he was simply realistic. He decided to drop the recurrent crusade for regulation, which had gained nothing but the antagonism of a powerful part of the Forest Service constituency. Just a week and a half after he had taken office, representatives of the Forest Service and of thirty-nine timber companies met in Atlanta to discuss their "mutual interests" in the South. Regional Forester Charles A. Connaughton told the gathering: "The U.S. Forest Service recognizes that public regulation is unattainable in the near future if it ever comes."[6]

At a stroke, McArdle ended an issue that prevented the Forest Service and industry from working together. Industry circles buzzed with speculation about what was really going on. Most leaders were cautious, and fear of the Forest Service's real objectives lingered for decades; nevertheless, a cooperative spirit emerged in the private sector. By January 1953, industry representatives had met with Secretary of Agriculture Ezra Taft Benson and pronounced McArdle a chief they could work with.[7]

The end of the regulation crusade did not leave the Forest Service's attention confined to internal programs. In 1952 it announced an ambitious, expanded version of the old forest survey, called the Timber Resources Review. The Service suggested that previous inventories were inaccurate and sought a comprehensive analysis of the nation's forests and of future demands for their products. Industry leaders feared an oblique stab at a renewed call for regulation. Several years of push and pull ensued, with the Forest Service keeping tight control of the project and with industry marshaling sophisticated commentary. The draft of the review's final report recommended a program for public and private lands, but thanks to industry influence, the final report proposed no program at all. It merely suggested that although there was no immediate danger of a "timber famine," there was no surplus of forests. It projected enormously increasing demands for forest products and implied that it would be advisable to pursue intensive management to keep up production.[8]

THE STRUGGLE OVER MULTIPLE USE

The attempt to develop a grand scheme for all the nation's forests came to naught, and the place of the national forests within the larger timber economy remained undefined. The time for expanding the National Forest System had ended with the election of President Eisenhower. McArdle announced to the Forest Service in 1953: "The Secretary and Assistant Secretary are both strongly—very strongly—opposed to further enlargement of Federal land holdings." They went so far as to place restrictions on even minor acquisitions, such as exchanges of land for timber. Throughout the 1950s, all attempts to define a national policy on the ownership of timberland ran afoul of bitter disputes over the proper extent of public ownership.[9]

The purposes and policies of the national forests seemed open for definition. "Aside from relatively small areas having special value for scenic or other purposes, the national forests are managed for timber production," Christopher M. Granger told a congressman in 1948. That was a voice from a simpler age. The policy of the government "is to understand the problems of every special group in the country, but never to use the resources of this country to favor any group at the expense of others," Eisenhower told the Fourth American Forestry Congress in 1953. That was a simple way of saying that things were no longer going to be so simple.[10]

"Multiple use" had been around for a long time. The 1897 forest-reserve legislation had acknowledged timber and water as their purposes. The national forests served other interests, however. Grazing was the most important until the 1940s, while mining and various forms of public recreation had required attention from the earliest days of the Forest Service. Since the 1920s, the Service had also shown important attention to wildlife and aquatics and had developed a formal, if unsystematic, program of designated wildernesses and primitive areas. Each activity required administrative response to public demand or to simple realities. As late as the 1950s, however, there was no system to accommodate conflicting interests.

The Forest Service was no stranger to conflict, but this had usually occurred in well-defined arenas, with allies and opponents easily identifiable, as in the long-standing feud with the timber industry. A legacy of vicious struggles over regulation had sorely weakened the Service's public position. It could no longer count upon the unquestioning support of even the American Forestry Association or the Society of American Foresters. Many other public groups questioned the agency's motives or judgment.

Ironically, it was the Forest Service's emphasis on timber that required a foundation for multiple-use management that was more concrete than increasingly frequent allusions to Pinchot's evocation of the "greatest good of the greatest number." Timber had always been important. Philosophically, national-forest timber was the future salvation of the nation, to become available when other supplies ran out. Pragmatically, sales receipts justified the existence of the forests and earned notice for foresters who produced them.

Until World War II, however, timber sales by the Forest Service were small change compared with the larger timber economy. Demand for the "allowable cut" suddenly increased during the 1940s. Access roads were built in order to open more areas to harvesting. When wartime restrictions ended, the general public climbed into its collective cars and entered the national forests in growing numbers, discovering that access for loggers could be access for others as well.

Timber had not only been the most important object of national forest management in the philosophy of the foresters; it had also been very demanding of time and resources. Some attention had been given to recreation and other values, but as long as the numbers and requirements of recreationists were comparatively small, recreation ranked below timber on the Forest Service's scale of priorities. With the public steadily increasing its demand for recreation after the war, a rearrangement of priorities seemed in order. With demand for timber also increasing, it remained to be seen whether a new scale of priorities could be developed.

Formal consideration of the concept of multiple use began in the Society of American Foresters during the war. In 1947 the society produced a document calling for multiple use as a basis of policy to ensure "adequate recognition of all resources and benefits," including recreation. William B. Greeley, however, demanded that such a policy be restricted to public lands and that it not be imposed on private enterprise.[11]

When the American Forestry Congress in 1953 included a session on multiple use, the Forest Service sent Assistant Chief Edward P. Cliff to state the Service's position. Cliff's career had included staff and line positions that involved range and wildlife management, fire control, timber management, wilderness, and public relations—multiple use personified. He told his audience that the two basic principles of national-forest management were multiple use and sustained yield. In combination, they meant coordinated management for the maximum "continuous" yield of products and services. Multiple use did not mean that every use was possible on every acre of ground, however: "Timber production is given priority over other uses on the most important areas of commercial forest land, with recreation, livestock grazing, and wildlife being integrated as fully as possible without undue interference with the dominant use." He also observed that increasing population and demands were making it more and more difficult to serve all interests together. "It must be remembered that every citizen of this country shares in the ownership of the national forests and has a right to fair consideration of his needs and desires in his use of them . . . but it certainly doesn't make multiple use any easier."[12]

Public pronouncements on multiple use often bordered on the platitudinous. To forester Paul A. Herbert of the National Wildlife Federation, talk of multiple use was a lot of nonsense if it was to be left to foresters alone. "Despite the general acceptance of the multiple-use principle, young men who elect to enter this profession are still being trained for wood production," he observed. Of twenty-seven forestry schools at that time, only one required a course in water conservation; three, in recreation; and eleven, in wildlife management. Such educations, he maintained, produced people who were "entirely inadequate" to manage public forests in which multiple use was the objective. He blamed the Forest Service, the Society of American Foresters, and the schools together for forestry's wooden narrow-mindedness.[13]

Foresters, of course, ran the Forest Service. Talk of multiple use notwithstanding, timber came first on their list. A proposed multiple-use plan for forests in the Sierra Nevada said: "Forage and wildlife are secondary to timber on timber producing lands. On sub-marginal sites no attempt will be made to raise a second crop of timber after the harvest cut. On the east side, it *may* be desirable, where grazing is to be the primary use, to have 100% [logging slash] disposal."[14]

The Forest Service was bound for collision with some of its oldest friends by the mid 1950s. That was the result of several things. For decades the Service had propagandized against "destructive" logging, leaving a definite impression on the minds of the general public. As timber harvests increased on the national forests, recreationists, who were looking for camping, hiking, fishing, or just scenery, encountered what appeared to be the very thing the Forest Service had railed against. Many people felt deceived. When timber harvests extended into virgin stands that were important to groups such as the increasingly active Sierra Club, organized outrage became a fact of public life. And when the Service and big industry suddenly ceased to say nasty things about each other in public, many people grew suspicious. Although the Forest Service maintained that "sanitation salvage" logging (the removal of hazardous or diseased trees) was compatible with recreation, it learned the hard way that some people wanted no part of any logging in recreational areas, no matter how necessary foresters thought it to be.[15]

Guardians of wildlife were swift to act, but the Forest Service was much slower to react. When a multiple-use law was first proposed in Congress in 1955, Leo V. Bodine of the National Lumber Manufacturers Association held a meeting with Chief McArdle and Ira J. Mason. Regarding the legislative proposal, McArdle told Bodine: "Just between you and me, I don't like it." McArdle dismissed the measure's proposed multiple-use advisory council as "just window dressing."[16] Against that attitude, the first multiple-use bill went nowhere. It remained in the docket, however.

The Forest Service's public-relations position was not improved when the National Lumber Manufacturers Association adopted a policy statement on multiple use that echoed the Service's emphasis on increased material production. According to the statement, "The lumber industry has long recognized the multiple use principle in the management of its forest lands. It also realizes that in the application of this principle certain priorities of use must be observed." Regarding public lands, "All national forest land is to be devoted to its most productive use for the permanent good of the whole people." There could be little doubt what a lumbermen's association regarded as "most productive."[17]

When the Forest Service ceased to castigate the timber industry during the 1950s, it lost an easily identifiable cause. The industry could no longer be branded with "cut-out-and-get-out" logging. It was not perfect, but it was becoming more and more efficient—far more so than the national forests, when it came to timber production per acre. Industry had improved its public standing considerably. It had also gained influence in Congress. Furthermore, cooperation with industry was essential if the planned development of the nation's timber resources was to be achieved.

To replace industry as a jousting partner, the Forest Service found itself facing a congeries of groups favoring wildlife, recreation, and wilderness. Their cause had galvanized over a failed attempt to flood Dinosaur National Monument. The Forest Service had had nothing to do with that proposal, but it found itself caught up in a general distrust of land-management agencies, a "land grab/giveaway" uproar that had been developing in several controversies since the late 1940s.

The Forest Service dismissed all critics, much as it had in the debates over sustained-yield units. Proponents of wilderness, especially, were scorned as "preservationists" and "protectionists," devoted to "locking up" valuable resources. Simple-minded epithets aside, it was understandable that the proponents of multiple use looked askance at any emphasis on wilderness. Wilderness seemed to deny the sort of balance that multiple use promised, especially if it extended to commercially valuable areas. To many recreationists, however, wilderness became the issue.[18]

The "locking up" shibboleth reflected a more palpable fear in the Forest Service—the National Park Service. The Forest Service had long recognized that the Park Service could be a formidable competitor in public-land recreation. The controversy over Olympic National Park during the 1930s showed how serious the effects could be if the public were to decide that the Forest Service could not be trusted to protect recreational values. The Park Service had an exceedingly energetic director during the 1950s, Conrad Wirth. Although his agency faced the same difficulties as the larger Forest Service—increasing demands upon inadequate staff and budget—he developed in 1956 an imaginative and congressionally rewarded solution. Mission 66 was a ten-year program to upgrade the physical plant of the National Park System. The plan also called for the expansion of the system, and it was inevitable that part of its growth would come at the expense of the national forests. In 1957 the Forest Service put forth its own version of Mission 66, Operation Outdoors, but it would not answer all fears.[19]

Assailed on all sides, defended by few, the Forest Service was in an awkward position. Bureaucratic self-preservation is among nature's strongest instincts, however. "Multiple use" was dusted off and given a new appearance. The Forest Service might become a moderate middle ground between the extremes of "total use" of timber, demanded by industry, and of "locking up" in the park system, demanded by others. The Service muted its old philosophy that it was the national defender against timber famine and pulled back. It would no longer exist primarily to cut timber on a sustained-yield basis; it would balance that with watershed, recreation, wildlife, and other values, all according to its sustained-yield principles. It could redefine its mission, somewhere between exploitation and preservation, and be all things to all people. But whether the old-line foresters in the Service itself could make such an adjustment remained to be demonstrated.

In 1956, Senator Hubert H. Humphrey introduced a bill to protect the multiple resources of the national forests, listing them as range, timber, water, minerals, and wildlife. He introduced another bill for the protection of wilderness. Sentiment for both bills increased in Congress over the next couple of years. Proponents of multiple use generally opposed the wilderness legislation, but the two issues became linked in the minds of the conservation groups most in favor of recreation. Compromise was inevitable.[20]

The Forest Service's leadership was divided on the multiple-use legislation. Even those who favored it advised caution, fearful of the consequences if the legislation should fail to pass. By 1958, McArdle had decided that the law was the way to go, however. His people, typically, lined up with him and presented a united front. When the Park Service attacked the multiple-use bill as an assault on its own ambitions, the Forest Service stepped up its campaign, with a great deal of emphasis on recreation.[21]

Edward P. Cliff became a principal advocate of the multiple-use legislation, at least before the public (he had misgivings at the time, although afterwards he became a true believer). Portraying the Forest Service's dilemma as one in which various parties were struggling over the forests to protect their own interests, he acknowledged that it appeared to be a series of conflicts between forest users and the Forest Service. "Basically, however, the conflicts are between the different groups of users, with the Forest Service in the position of a referee who must make the decisions and enforce the rules of the game."[22]

The multiple-use bill had originated with conservation organizations, seeking to protect recreation and wildlife values from the growing timber-management program. Virtually all conservation groups supported the measure, the notable exception being the Sierra Club, which feared that the bill would endanger the wilderness legislation. The timber industry opposed multiple use as a possible threat to its own interests. McArdle himself told an industry group that "during the last 10 or 15 years, greater emphasis has been given to increasing the timber harvest on the national forests than to some of the other resource-management activities." Unless a balance could be achieved among competing interests, he warned his audience, the reaction of the public could be serious indeed for the timber industry. He told them that multiple use was the wave of the future, "the only possible way to strike a livable balance between timber production and these other national forest uses."[23]

Out in California the regional office produced a pamphlet explaining multiple use as "a concept of land use rather than a system or method" and saying that while multiple use required great skill, it was the only way to serve competing demands. Cliff told a conference of loggers: "Multiple use is on trial today. With increasing use of all resources and more people ready to criticize things they don't like, multiple use will continue to be on trial. . . . Foresters and loggers have a big job to convince the public that timber harvesting can be fitted into a sound program of multiple land use." He advised his listeners to help to "make multiple use work and keep timber production in its proper place in the multiple-use picture on the national forests."[24]

In 1959 the Department of Agriculture forwarded to Congress the "Program for the National Forests," an attempt to obtain balanced budgeting for an expanded program in the multiple-use spirit. "Operation Multiple Use," Cliff called it, promised "more wood for the nation," "better hunting and fishing," "more and better water," "more outdoor recreation for more millions," and "better range, better grazing." By 1960 the Washington office was thoroughly committed to multiple-use legislation; for about a year, Cliff spoke often, at a variety of locations, exclusively on that subject, including an address to the Fifth World Forestry Conference in Seattle in September 1960, where he declared: "We in the Forest Service believe that multiple use offers the best hope of stretching the resources on a shrinking land base to meet the expanding needs of our people."[25]

The Multiple Use–Sustained Yield Act became law on 12 June 1960. Without assigning priorities, it stated that the national forests were to serve five major interests—timber, wildlife, range, water, and outdoor recreation. Each was to be managed according to the principle of sustained yield (the Forest Service's, not David T. Mason's)—that is, at the maximum rate of extraction that could be sustained without depletion of the resource. While all interests were to be served, they need not all be served in the same place.[26]

Only time would tell whether the legislation would effectively redefine the mission or reorder the priorities of the Forest Service. Much would depend upon whether Congress would be as willing to pay for recreation management as for such programs as timber and grazing, which produced income for the Treasury. The timber industry did not like the act, but had to live with it; the industry was bound to guard its own stake in the national forests.

Ultimately, whether the Multiple Use Act meant anything would depend upon the Forest Service. The act had come after phenomenal increases in recreational use of the national forests and at a time when vigorous new recreation programs were taking shape. It had also followed two decades of equally remarkable growth in the timber sales on the forests. As Herbert had observed in 1953, foresters were oriented toward timber from the time they left school. Increasing sales during the 1950s reinforced that orientation. Moreover, the Washington office had made little effort to explain the new legislation to people in the field; some of them at first resisted its implications. The Multiple-Use Act, therefore, asked an organization that was composed of people inclined to focus on one thing to think equally about several other things, even if at the expense of their principal object. It was a significant challenge, made greater by the strength of timber's grip on the Service's consciousness.

FOCUS ON TIMBER

"The two basic principles which govern the management of the national forests are multiple use and sustained yield," Edward P. Cliff told the Fourth American Forestry Congress in 1953.[27] Although he became a principal salesman of multiple use later in the decade, throughout the 1950s Cliff was noticed mostly as a promoter of sustained yield. As he put it, sustained yield was taken to its logical conclusion to mean increasing harvest to reach the maximum "allowable cut," or the point at which extraction was equaled by replacement.

In 1956, Cliff said that "the overall timber policy of the Forest Service is to build up the annual cut to the full [amount] allowable under sustained yield management in each working circle." He observed that the estimated allowable cut on the national forests (exclusive of Alaska) had recently been revised upward, from 6.5 billion board feet per year to 7.6 billion feet, and he predicted an allowable 9.5 billion feet by 1960.

Cliff said that current timber-management programs had four objects. One was the completion of inventories and management plans for all working circles, which with increased funding could be done in six years. That, he expected, would produce increases in allowable cut, because of advances in estimating techniques, because road building and increased stumpage values had "revolutionized" concepts of what was commercially operable timber, and because of improvements in timber utilization (using more of the tree, defective trees, more species, and so on). The second emphasis was on salvage logging, the harvesting of scattered dead and dying trees, which theretofore had been hampered by lack of access and marking. The third was "intensive forest management," or the control of insects, the promotion of reforestation, the thinning of second-growth stands, the use of improved varieties of seeds, and the like. He said that the fourth and most important objective of the Forest Service was "to close the remaining gap between actual and allowable cut." That would be achieved by improved accessibility and by clearing rights of way over private land, if necessary by condemnation. Cliff also wished to see the extension of Forest Service timber management over mining claims, the resolution of questions over proposed dams, and more housing for timber-sale personnel. He concluded by observing that in the previous decade, harvests on the national forests had increased from 2.60 billion to 7.25 billion board feet, with a ninefold increase in the money involved, to about $100 million per year. "We want to keep our policies and procedures growing in line with these increasing opportunities," he said.[28]

Cliff never called directly for increasing the annual allowable cut on the national forests, although he was fond of predicting that better information would lead to an increase. Raising sales to reach the allowable cut was his most recurrent theme, which he connected invariably to the development of access roads. In 1957, after his first five years as assistant chief for National Forest Resources Management, Cliff's staff listed the allowable-cut crusade as his greatest achievement: "The harvest of national forest timber has increased from 4½ billion board feet in F.Y. 1952 to 7 billion board feet for the year ending June 30,

1957. The present harvest is more in line with the allowable cut than it has ever been."[29]

Cliff supported his campaign with bright predictions about the future productive potential of the national forests. "For example," he told a congressional committee in 1957, "by the year 2000 the national forests should annually produce 24 billion board feet of timber if they are to carry their full share of the job of producing wood to meet the Nation's anticipated requirements by that time." To that end, he said, the government and the purchasers of national-forest timber were building about 2,800 miles of new roads each year to get the timber out.[30]

Multiple use reappeared in Cliff's discussions of sustained yield in 1958: "In combination these principles mean coordinated management for the maximum *continuous* yield of beneficial products and services from the national forests." To another congressional committee he avowed that "the objective of the Forest Service is to cut the full allowable rate for each of the 422 working circles." He estimated the allowable cut in 1957 (excluding Alaska) at 9.89 billion board feet and bemoaned the actual 1956 cut of 6.91 billion feet. Access roads, he said, were the way to bring the two figures together.[31] Cliff argued ceaselessly for more roads to increase national forest harvests. People need not fear for the forests, he said, as the Forest Service "recognizes that forest resources are renewable and can last forever if intensively managed."[32]

TIMBER MYTHOLOGY

The Forest Service, McArdle said in 1958, "believes" that the nation had a permanent expanding market for forest products. Estimating that 24 billion board feet of timber would be required annually from the national forests by the year 2000, McArdle deplored the limitations that held the harvest to about 7 billion feet. "These volume objectives ought to dispel any doubts as to which way the Forest Service faces on full-scale timber production and utilization of timber on the national forests." To meet the objectives would require the delivery of "the full allowable cut from practically every working circle."[33]

The Forest Service's leaders actually believed that the nation's demand for wood would continue to grow for a long time into the future. Contained in that belief was the assumption that the national forests were the chief producers of wood for the future, a legacy of longstanding doubts that the private sector would ever reforest the nonfederal landscape, once it had taken its cut. In fact, however, the belief was little more than a belief. That it was held strongly did not change the fact that it did not reflect what was already happening in the national timber economy. Sales of national-forest timber continued to increase until the mid 1960s, when they leveled off at between 11 and 12 billion feet per year and remained there.[34] There were, of course, budgetary and political developments that helped to end the increases in sales volume, but they alone cannot explain the sudden end in sales growth.

Demand for national-forest timber was not offset by unpredicted increases in private production. The latter did increase, but demands for it paralleled, rather than supplanted, increasing demands for national-forest timber. Demand, in fact, fell off, in particular for lumber. The total of domestic lumber production from all sources leveled off at about 38 billion board feet in 1950, fell slightly over the next five years, peaked at 38.2 billion feet, then fell off to 32.7 billion feet in 1957. There was another peak of 37.2 billion feet in 1959, after which production fell off again. By the mid 1960s, production had leveled off at between 34 and 36 billion feet, where it remained thereafter.[35]

There has been, in fact, a general decline in the per capita consumption of forest products for decades, with the most dramatic decreases since the mid 1950s being in lumber. That decline has been enough to offset increased consumption attributable to the growth in the country's general population, let alone old notions that per capita consumption must increase with the development of society. The total apparent consumption of forest products in the United States, excluding fuel wood but including imports, was about 9.91 billion cubic feet in 1950 and about 10.49 billion cubic feet in 1955. There have been annual fluctuations since then, but there has not been enormous growth. After a decade of relative decline, then resurgence, consumption stood at 11.93 billion cubic feet in 1965. Total consumption in 1970 was 12.18 billion cubic feet—in other words, it had been fairly level for several years.

Despite clamoring in the marketplace for greater access to national-forest timber, there was no enduring upward pressure on market demand sufficient to require projected harvests of 24 billion board feet by the end of the century. Without such increases, there was no compensating pressure on other sources, most notably imports, which actually experienced modest decreases. Net imports of all timber products, excluding fuel wood, stood at 1.38 billion cubic feet in 1950 and, with minor fluctuations, declined slightly thereafter, standing at 1.06 billion cubic feet in 1970. In contrast, the net exports of logs have shown dramatic and continuing increases since 1950, when they amounted to 10 million cubic feet, against 45 million feet imported. Exports in 1970 amounted to 430 million cubic feet, offsetting a mere 25 million feet of log imports. Far from facing a timber famine, the United States proved to be increasingly able to supply itself and others with forest products.

A closer look at the consumption of various types of forest products is more instructive. The consumption of lumber remained fairly steady after 1950, between 5.5 and 6.4 billion cubic feet per year. The consumption of plywood and veneer increased regularly, however; and the consumption of pulp products grew almost as dramatically. While the general national consumption of forest products remained relatively stable during the two decades after 1950, there was a strong shift in consumption away from lumber and toward a proportionately greater use of plywood/veneer and pulpwood products. The latter, in other words, were increasingly substituted for the former, which suggests that quality lumber became more valuable, probably by reason of scarcity. That was reflected in imports, which between 1950 and 1970 declined greatly for pulpwood but increased from 5 million to 155 million cubic feet for plywood/veneer—and from 455 million to 755 million cubic feet for lumber. (It should be recognized that exports of lumber also increased significantly during the same period.)

As early as 1953, J. Alfred Hall, director of the Forest Products Laboratory, warned his forester colleagues that their assumptions about the production and use of the national wood supply were flawed. To begin with, he urged more thoughtful and responsive timber management, to answer the nation's true and evolving needs. Timber crops, he said, ought to be more commonly committed to management for the production of multiple products, such as veneer logs, saw logs, poles, piling, pulpwood, and so on. He observed: "This is the direct antithesis of some of the single-product management of the past and as our facilities for conversion of all species to useful fibre products become more widespread, it would seem to be quite logical to anticipate an extension of management stands of mixed species [in contrast to a] monoculture that in many cases has already proven unsatisfactory."

Hall also thought that it was "the task of research to anticipate developments in utilization fields and orient management policies in such directions as to do two things—preserve and enhance the growth capacity of the forest, and satisfy our national needs in wood products." He then put his finger on the fallacy of assuming that there would be indefinite increases in the annual cut: "I want to emphasize my belief that, in such [intensive] management, special attention will be given to the production of wood of high quality grown from pruned trees of even-growth rate. Manufacturing and consumption requirements in the future, I believe, will not tolerate the quality of much of the wood we are now using from our second-growth forests."[36]

Hall's position gave him the outlook of a user of forest products. To the forester, wood was accounted for sufficiently by reckoning volumes, market value, species demands, and the like. To the user, wood was a material with which to make things. In the case of the most important species, the best wood for construction usually comes from old-growth trees.[37]

Demand for national-forest timber increased during the 1950s because it represented the largest remaining ownership of virgin timber, regardless of species. Private ownerships had been sorely depleted during the preceding decades. By the 1950s, their production of second-growth timber was a subject of widespread pride, but that did not satisfy requirements for high-quality old-growth timber to support the construction boom. Industry therefore turned increasingly to the national forests for supplies of old-growth timber, and the Forest Service, in the interest of timber management and sustained yield, was happy to oblige so far as its budget would allow.

The two interests coalesced for a period, but their outlooks were different. From the construction standpoint, increasing harvests in old national-forest stands solved a problem of immediate shortages. Industry probably realized that the immediate extraction of old-growth timber premised the ultimate exhaustion of all older stands. Increasingly during and after the 1950s, the construction industry and other users of forest products revised size and grading standards, turned more and more to materials that substituted for wood, and developed composition products that stretched or conserved timber and permitted the use of young timber and logging wastes that were not otherwise suitable for construction. That, in part, was an adaptation to the announced Forest Service policy of forest conversion (from "overmature" timber to second growth). It was mostly an acknowledgment that eventually—and thanks to industry's clamoring demand for the stuff—the nation's old timber would no longer be commercially available. Accordingly, the increased demand for pulpwood and small products offset declines in lumber just enough to keep the total national-forest sales from declining during the 1960s but not enough to sustain the former rates of increase.

The response of the federal foresters was rather different. Old-growth timber stands, they thought, were "unproductive," because they did not increase their standing volume at the same rate as young timber. The mortality of old growth offset its "increment," so that its static volume did not contribute to allowable cut, whereas young-growth timber increased in volume rapidly. Old growth was, therefore, not a precious high-quality lumber resource so much as it was an obstacle to achieving higher annual production of timber volume. The current high demand for old growth permitted both accelerated conversion to second growth and maximum rates of sustained yield, so far as the foresters were concerned.

At a time (the 1950s) when it appeared that the Forest Service and the principal forest-products industries were most in agreement about what needed to be done, they were in fact speaking quite different languages. The Forest Service had taken its old objective of sustained yield to its logical conclusion. Very moderate and selective harvests in a virgin forest could in fact be maintained indefinitely, because the natural replacement of timber would match or exceed extraction; second growth would be old before it was harvested. With demand low in the early years, that was what sustained yield meant in practical terms.

The accelerated demands of the 1940s and 1950s offered an opportunity to raise sustained yield to a maximum level—when the greatest volume of natural replacement matched the greatest volume of extraction. When ''overmature'' timber was removed and when all the commercial forests were of younger, faster-growing stock, forests would be more ''productive,'' and harvests could be increased. The Forest Service believed that its version of sustained yield would perpetuate a resource that was vital to industry, but an accelerated drive for sustained yield really meant replacing one resource, old-growth timber, with another.

Old-growth timber was ''renewable'' only if one were willing to wait centuries for the renewal. Volume measurements were deceptively simple but, to users of wood, could not hide the differences between young wood and old. In practical terms, when industry found old growth progressively more expensive, it learned to do without. The rate of increase in consumption of national-forest timber moderated, and sales leveled off when such things as concrete blocks and fiberboard-photoprint paneling became more common than real wood in buildings. Meanwhile, increasing lumber imports helped to offset the progressive scarcity of domestic old growth, of which, for some purposes, fiber products (from second growth) and other substitutes were not good. The increasing consumption of national-forest timber, focused from the buyer's standpoint an old growth first, was bound to level off with the conversion to second growth.

There were other segments of the public who had other views of the forest, also greatly divergent from those of foresters. If dedicated foresters viewed old growth as ''overmature'' or unproductive timber standing in the way of sustained yield, other people regarded it as ''virgin'' forest, a home to wildlife, nice to look at, a place for recreation, or—most important politically—the repository of the social benefits emanating from ''wilderness.'' Harvesting of ''overmature'' timber, which was desirable from the lumberman's or the forester's standpoint, was an abomination to the lover of the "virgin" forest. Until the recreationist and the conservationist could come to believe that their sentiments were being accorded attention comparable to those of the timberman, they were likely to raise a cry. By the early 1960s they had begun to raise many cries.

Multiple use emerged late in the 1950s as an attempt to deal with the realities of the new age—one of which was that the clients of the national forest were many. It was a product of improvization during a period when a hard-pressed timber industry was pulling at the Forest Service, on the one hand, to increase sales, while on the other hand, recreational and wilderness proponents were demanding, first, more attention and, ultimately, some limits on the disruption caused by timber harvesting. During the 1950s, national-forest timber management was pushed heavily in the direction of the maximum allowable cut, and in the 1960s it ran headlong into forces that wanted to reduce the volume—and the area—of the cut. Multiple *use* ran afoul of multiple *demands*.

TOWARD MAXIMUM SUSTAINED YIELD

The Forest Service's strong momentum toward the maximum sustained yield of timber, despite other considerations, was demonstrated in a number of management plans adopted in the 1950s. Policies on the Big Quilcene Watershed of the Olympic National Forest, which were adopted tentatively in 1955, were very interesting. The city of Port Townsend, Washington, expressed anxiety over the careful management of the Big Quilcene Watershed, the source of its water supply. The Service reassured the community that it would "propose nothing in the way of management that will adversely affect the amount and purity of the water supply." Over half of the watershed, however, was covered in "commercial timber types," capable of producing 9.5 million board feet per year "indefinitely." The policy statement said:

> Since the watershed contains such a high volume of mature timber, it is essential that harvesting be started soon to avoid the necessity of a larger yearly cut at some later date when mature timber in the Quilcene Working Circle becomes nearly exhausted. From a timber management standpoint it is desirable to make a heavy cut on the watershed in the near future. However, the Forest Service recognizes the desirability of keeping the area logged each year to the minimum necessary to cut the annual sustained yield of 9,500,000 board feet. On this basis it is now planned to cut each year only the annual increment that is produced on the Watershed itself.

Timber would be harvested by clearcutting. To reassure the community, however, the Forest Service said that the cutting patches would be limited to no more than thirty acres each (the normal practice in the area was then a maximum of eighty acres per patch) and that great care would be shown in logging practices. As soon as slash was disposed of, each cutover area would be artificially planted.

The telling point was that the Forest Service believed old timber simply must be harvested. To be sure, the job could be done most carefully on the foresters' terms, and there were arguments that fire and parasites could imperil a watershed. But the main point seemed to be that old growth simply had to go, in favor of second growth, and that all other considerations were irrelevant. To underscore that point, the report ended: ''In conclusion, it should be stressed that timber harvesting on the watershed is inevitable and to delay longer means more extensive and severe cutting when the completion of the rotation in the Working Circle eventually forces logging activity onto the watershed.''[38]

A management plan for the Snoqualmie National Forest, adopted in 1956 and distributed as a model for other forests in the Pacific Northwest, had two chief objectives. One was to put lands ''to their highest use.'' The other was to harvest a ''maximum continuous volume'' of timber each year, while keeping all lands productive. That was to start with harvests of 40 to 70 percent of allowable cut, with a goal of reaching full levels of allowable cut within five years. The Snoqualmie plan required the selection of the ''best silvicultural system,'' but emphasis went to the fastest possible development of roads to permit salvage and other sanitation and improvement cuttings. The forest's Douglas-fir would be harvested in clearcuts ranging from ten to one hundred acres, with sizes subject to change. The purposes of maintaining the maximum allowable cut were to contribute a ''full share to the economy of the region and convert stagnant stands to a growing condition.''[39]

The Forest Service was already a decentralized organization, but in 1955 it passed the widest possible authority for technical decisions down to the lowest level, in order to accelerate timber sales. The Service *Manual*'s section on ''Timber Use Policies'' was amended to begin: ''There is so much variation in the characteristics of timber species and the condition and age of timber stands within the national forests that no generalization as to cutting methods can be classified as either reproduction cutting or intermediate cutting.'' The only accountability expected of the local level was the statement ''When a reproduction cut is made the ranger and forest supervisor are, in effect, under contract to reproduce the area cut over as promptly as practicable.''[40] The Forest Service set itself up for some rude shocks by disavowing any central responsibility for technical decisions at the local level. In later years, when parties aggrieved at what they regarded as improprieties in resources management in one or another national forest took their grievances to court, it would be the secretary of agriculture and the chief of the Forest Service, not the local ranger, who would be listed first among the defendants.[41]

INFORMATION AND EDUCATION

Growing public attention to what was happening on the national forests caused Forest Service leaders to conduct vigorous programs of "information and education," a new name for "forestry as propaganda." In the Great Lakes States, where second-growth harvesting was coming into its own, Cliff told a group that demands for recreation, iron ore, and better management were "problems to be reconciled" but that the future of the region lay in timber. "Of this I am certain, if you use the axe and the saw wisely, the forests of the Lake States will occupy an even more important and material place in the expanding economy of our country."[42] Cliff told congressional committees that the national forests were now "big business," and he asked for the budgets for roads, employee housing, and other wherewithal to make the business bigger in order that the 422 working circles might contribute that 24 billion board feet by the end of the century. Industry was dissatisfied at the rate of increase in sales; spokesman George Craig claimed that "our national forests are yielding one-fifth less than good management indicates is possible on a conservative basis."[43]

Chief McArdle did not like that kind of talk, so he asked for an end to "bickering" between the Forest Service and the private sector. Pointing out industry's dependence on national forests on the West Coast and observing the increasing numbers and influence of recreationists, he said that "some view with alarm the rate at which roads and timber cutting are invading the upper country." Unless the Service and industry coordinated their actions, he predicted, circumstances would "surely lead to public demands for greater restrictions and reservations for watershed protection and public use." In order to avert such perils, he asked industry's cooperation in improving erosion control, reducing damage to residual zones, reducing breakage of timber, improving wood utilization, and improving the cleanup of debris; he promised that the Forest Service would account for the increased costs of operations in its appraisals.[44]

Cliff, meanwhile, reassured an increasingly restive public that the Forest Service knew what it was doing in timber harvests. In a 1958 article on "The Care and Use of National Forests," he informed his readers that harvesting was a very technical matter, with several methods having been developed through experiment and trial.[45] The wilderness movement raised public concern about timber harvests on the national forests, a definite worry for Forest Service leadership; therefore Cliff discussed wilderness often. The Forest Service was a pioneer in wilderness, he said, and he underscored his own commitment to the idea. Cliff could go only so far, given the emphasis on the economic contributions of national-forest timber. "The real question is how much of the land needed by man to satisfy his hunger for the bare essentials of life—food, water, and shelter—is the Nation justified in removing from commodity use?" In any event, he advised industry to avoid "horrible examples where multiple use is not working" and to recognize that "foresters and the wood-using industry have not yet convinced the general public that logging is good land use." When push came to shove, however, the Forest Service's materialism was clear in Cliff's words in 1961: "The first and probably most important fact which must be faced is that trees and lesser vegetation are organisms; thus temporary. Each has its optimum life for utilization and if not utilized dies."[46]

Timber management had to make some concessions to other interests, however. By 1962, forests in the Northwest were practicing less clearcutting than formerly because of "regeneration problems," while such clearcutting as continued was being modified. Cliff called for more careful supervision, in order to avoid conflict, of "patch clearcutting near recreation areas wherein natural vistas are modified by silvicultural practices. Some recreationists vigorously oppose any form of timber cutting where the natural landscape is changed."[47]

Timber-sales procedures underwent some review during the 1950s, but they emerged essentially as they were before. In 1952 the Western Pine Association asked for a number of technical changes in appraisals and sales that would be favorable to purchasers. The next year, Cliff suggested that despite the sustained-yield-unit legislation, competitive bidding was most beneficial to the industry and communities. Large-volume and long-term sales, he said, while they had their merits, benefited only the large operators. In 1956 he announced a new sales-contract form, developed after wide review in industry and the Forest Service, which he said did not "signify any major shift in Forest Service policy or procedure."[48] Spokesmen for industry continued to maintain that the Forest Service's appraisal procedures put the purchasers at a disadvantage. Nonetheless, they bought the timber faster than they could cut it.[49] As public concern about timber harvests increased, Cliff felt called upon to point out that the majority of timber sales were small. Rather than turning the national forests over to big industry, he said, the Forest Service served the small purchaser. Over 90 percent of the thirty thousand sales each year, he reported, involved less than $2 million in appraised value. Only about a quarter of the volume that was sold went in sales of more than a million dollars, or 25 million board feet. Accordingly, Cliff spent much of the decade campaigning for road construction, and in that he had the strong support of industry, especially the small operators.[50]

INSECTS, DEER, AND OTHER WILDLIFE

During the 1950s the Forest Service dramatically increased its use of chemicals to control insects. Early in the program, studies suggested that DDT, the principal chemical in use, had no effects on fish or wildlife, except insofar as they might suffer from having fewer insects around for food. By 1962, however, the Service was dismayed to see that very often, despite the growing use of sprays, insect populations were increasing in target areas.[51]

Wildlife was one of the resources that figured in timber-management planning. The Washington office complained when a timber plan neglected the subject. One Region thus learned that a plan ''should have something more positive to say about wildlife management and not limit itself almost entirely to a discussion of wildlife damage to timber resources.''[52]

Wildlife received special notice for several reasons. For one, sportsmen had long been organized and, in many states, were powers to be reckoned with (Forest Service wildlife management was conducted in concert with the states, under federal law); they could not be ignored. Furthermore, the most prevalent wildlife problem in the West and around the Great Lakes was an overpopulation of ungulates; there was little worry that timber harvests would reduce them. On the contrary, timber harvesting promoted the well-being of large browsing animals, by creating openings and by fostering the kinds of growth that they depended upon. That became an article of faith. ''In most forested areas any approved cutting that results in broken and mixed cover types is generally conceded to be beneficial to wildlife, as it improves the browse and cover,'' according to the Forest Service's *Manual*. In the Appalachian forests, in fact, a lot of timber was clearcut and left on the ground purely as a wildlife-habitat improvement.

That happy compatibility of harvesting and deer ultimately had some unexpected effects on the Forest Service. It worked so well that eventually, ''wildlife,'' to many foresters, meant deer and their cousins, the beneficiaries of harvesting. Although the Forest Service showed increasing attention to other animals—rare species of sparrows around the Great Lakes and turkeys in some parts of the East—during the 1960s the federal foresters came into conflict with some Appalachian sportsmen who viewed such creatures as turkeys, squirrels, and others as desirable wildlife. Many of those smaller beasts were affected by certain timber harvests very adversely.[53]

''Multiple use'' may have deluded the Forest Service into the belief that it would accommodate all interests in the national forests. But dividing the forests into five categories of ''uses'' was simplistic, and it could not loosen the grip that timber production held on the agency's outlook. It made it possible for the Service, without counting their numbers or weighing their views, to lump many elements of the public into a faceless mass of ''users''—recreationists or wilderness advocates—who could be tamed by ''information and education'' to see themselves as being no more entitled to their wants than were other ''users.'' It also encouraged such misperceptions as the casual translation of ''wildlife'' into deer and their relatives. But the people, like forest fauna, were exceedingly diverse, and their increasing demands on federal land managers could not easily be pigeonholed into a few labeled ''uses.'' Trying to appear as all things to everyone, the Forest Service was in danger of becoming nothing to no one.

CHAPTER SEVEN

From Multiple Use to Sustained Planning

I once heard a Senator tell a Regional Forester that he "did an excellent job of talking to the trees, but that he was not effective in talking to people." Our ranks are filled with highly competent professional employees. Most of them have to deal with people in addition to technical problems. However, they lack or have not developed the ability to be effective with people.

—Zane G. Smith (1968)

The Forest Service hosted the Fifth World Forestry Congress in Seattle in 1960, choosing multiple use as its theme. Delegates from around the world heard Chief Richard McArdle evoke Pinchot's call for the "greatest good of the greatest number in the long run" and say: "These instructions have constituted Forest Service doctrine from the beginning. They are the genesis of multiple use." It required careful planning, rather than happenstance, to produce success; but multiple-use management could reduce competition by preventing any one use from dominating others. "It offers balance," McArdle said.[1]

Those were fine words, but as became increasingly apparent over the next two decades, what the Forest Service said and what it really believed were not always the same. Timber remained the agency's principal focus, and "forestry as propaganda"—now dressed up as "information and education"—remained its principal way of dealing with the world. Seeking to deflect criticism without reexamining its own outlook, the Forest Service became increasingly subject to criticism—then to legislative and judicial action. Ultimately, its opponents nearly shut the timber program down until the Service was forced by circumstance to change its ways, if not its beliefs. But after the dust had settled, the Forest Service remained, shielded from critics by "multiple use" in refined form—a planning procedure without end.

TIMBER AND MULTIPLE USE

The Forest Service beat the drums for multiple use throughout the next two decades, using it as a shield against extreme demands from any one segment of the public. At the same time, the agency pressed ahead with revisions of timber-management plans to increase and attain the allowable cut. Roads extended into formerly inaccessible regions, inventories were updated, and cutting schedules were revised.[2]

Vague and all-embracing as the multiple-use idea was, it was also subtle. The concept's essential appeal was that the same area could provide for more than one use—timber, grazing, recreation—although practices would have to be adjusted to accommodate a mix. The easiest approach was to compartmentalize national forests according to "primary values." Following that principle, a tract that offered a lot of timber was perforce regarded as primarily valuable for timber and was managed accordingly. The Forest Service was reluctant to use the word, but where timber was present, timber management tended to be the "dominant" use. According to one document in 1962: "These plans delineate the primary or key resources for each land area. On areas with primary timber values, stands will be managed as intensively as possible with maximum volume and quality production the primary objective."[3]

Moreover, timber management was not to be excluded from areas dedicated to other purposes, so long as there was wood to be had. For years, Edward P. Cliff maintained that some kinds of timber harvesting were possible even in recreation areas. When he became chief in 1962, he put that principle into practice. Rejecting a plan for "near natural" management of certain zones of a California national forest, he declared that Forest Service policy called for a certain degree of harvesting in some scenic areas: "To maintain *all* parts of *all* scenic areas 'as nearly as possible in natural condition' is obviously impracticable and would be undesirable even from the standpoint of aesthetics." Vacating earlier instructions to maintain scenic areas in "undisturbed condition," he called for careful prescriptions to suit the circumstances. He also observed: "In recent years it has become quite obvious that 'near-natural' management is practically equivalent to no management at all. . . . It follows that it is not logical to authorize the establishment of virtually unmanaged areas of up to 100,000 acres."[4]

Cliff ordered that formulas for calculating annual cuts be revised according to the annual increment in "landscaping management areas," with consequent increases in allowable-cut figures. A 1962 report to the secretary of agriculture, calling for increased harvests in recreational areas, said "that there will be cutting on such areas and therefore, it is thought that the magnitude of future changes in overall allowable cut will not materially reduce the over-all allowable cut." It was determined further that "the cuts from landscape management areas are to be established separately and that recognition is given to the fact that realization of this portion of the allowable cut is dependent upon the ability of the Forest Service and industry to meet the special conditions of cutting on such areas.''[5]

Policies like those, however cautious, disturbed many recreationists, particularly proponents of wilderness. They believed that certain qualities might require the exclusion of development in some areas, even where commercially valuable timber was involved. The Forest Service had long been the pioneer in wilderness, but its designated areas tended to exclude accessible or valuable timber. They were located mostly in inaccessible alpine areas, where the scenery was outstanding but where there were few highly valued species of timber. Many highland trees were considered unmerchantable unless more-valued species were also present. With timber demands growing and with roads creeping into previously untouched areas, wilderness advocates feared that the Forest Service might retreat from the wilderness idea, pronouncements to the contrary notwithstanding. There was a growing perception that for the Forest Service, timber, where it existed, came first.

Forest Service participation in the activities of the Outdoor Recreation Resources Review Commission (ORRRC) during the late 1950s and early 1960s did not assuage such fears. ORRRC proposals emphasized developed recreation, the construction of roads, and other facilities not regarded as compatible with undeveloped wilderness. ORRRC recommendations resulted in the establishment of the Bureau of Outdoor Recreation in the Department of the Interior in 1962, a national program to acquire and develop recreational lands, and a fruitless attempt at a national plan for outdoor recreation.

Wilderness advocates wanted something else. The Sierra Club had withheld opposition to the Multiple Use–Sustained Yield Act in return for its acknowledgment that multiple use was not inconsistent with wilderness. That organization and others then sought legislative control of wilderness, founded on legislation first introduced by Senator Humphrey in 1956. The Forest Service opposed the wilderness measure until it gained reassurance that the bill would not counter the purposes of the Multiple Use Act, while the Park Service objected for reasons of its own. Neither position promoted confidence in the intentions of federal agencies. The agencies were joined in their objections by strong assertions from timber, mining, and other industries. That seemed to confirm fears on the other side that the agencies served commercial interests more willingly than they served the less organized general public. The latter viewed public lands primarily as places for recreation—and sometimes failed to distinguish between wilderness and other recreational areas. The result was the Wilderness Act of 1964, which removed existing national-forest wilderness areas from administrative discretion, provided for a ten-year period of congressional review of additional wilderness proposals, and established the Public Land Law Review Commission to examine all aspects of public-lands policy.[6]

The Wilderness Act reflected an absence of faith in multiple use or in the intentions of the Forest Service, whose feelings accordingly were bruised. The bureau's reactions to the legislation did little to help its standing with wilderness advocates. Although the Service, in obedience to the act, advanced quite a number of wilderness proposals over the following years, too often its public pronouncements appeared to be more negative than positive—describing the loss to the timber economy from wilderness designation, rather than advancing the values served by wilderness protection. Moreover, although other agencies developed wilderness proposals in similar landscapes, the Forest Service resisted the idea of wilderness in the East until supplementary legislation in 1975 required the Service to make serious efforts to designate wilderness in eastern national forests.[7]

Fascination with contributing timber to civilization underscored the Forest Service's uncertain attitude toward other values and affected its interpretation of multiple use. The timber-management plan for the Willamette National Forest in 1965 said: "Let it be understood, therefore, that all resources on this Forest are of equal importance in the basic concept of multiple-use management, and that no one resource shall be allowed to assume an over-riding position."

The plan, however, was couched in terms that measured all other uses against timber. Good timber management, according to the plan, was the "most useful tool for enhancing other uses." The following are examples:

> On timber management and scenery: "Clearcuts break the monotony of the scene, and deciduous brush in these areas furnish fall color and spring flowers for at least 10–15 years."
>
> On timber management and roads: "The roads built in connection with timber harvesting furnish access for more people to enjoy the use of other resources."
>
> On timber management and soils: The orderly harvest of old-growth timber "can reduce soil impacts from uproots."
>
> On timber management and recreation: "Openings also improve berry picking and are highly favorable to the development of large herds of deer and elk. This increases the potential for hunting."
>
> On timber management and multiple use: The timber manager "should direct his management of timber towards the use of timber harvesting as a tool in attaining true multiple-use management."[8]

It took quite a leap of logic to go from the foundation of multiple use—that timber was one of five equal uses of the national forests—to the belief that timber management was integral to every other use. The leap was made throughout most of the Forest Service. Where it was not, as in the Southwest Region, the Washington office was quick to complain: "One cannot help but feel that behind a facade of multiple use many people in the Region continue to be primarily concerned with grazing and recreation, often at the expense of other resources and uses. . . . Some honest self-examination is needed here." Washington wanted timber to be emphasized.[9]

Multiple use was conceived as a way to balance competing uses on the national forests. During the 1960s, curiously, the Forest Service was attacked for not actually applying multiple use. That allegation was not surprising when it came from recreationists who claimed that timber had priority in the Forest Service's scheme of things. But similar charges came from industry, which was restive about competition between its own interests and proponents of recreation and wilderness. When Regional Forester William D. Hurst of the Southwest Region attended the annual meeting of the Western Forestry and Conservation Association in 1967, he returned with a trenchant observation: "The climate with industry has apparently changed drastically in recent years with everyone using Multiple Use, as they see it, as a 'crutch' to support their program. It is, however, causing considerable modification of their previous militant stand for their own interests."[10]

The industry did modify its tactics during the late 1960s. Always strong supporters of funding for Forest Service timber programs, timber buyers were dissatisfied with the results. Seeking to expand its access to national forest timber, the industry prevailed upon supporters in Congress to introduce a bill entitled National Timber Supply Act. It would have created a revolving fund of timber-sale receipts, thus greatly increasing monies for timber activities. That flat disavowal of multiple use attracted widespread and fatal opposition, from the Forest Service and almost everyone else. A subsequent attempt by President Richard M. Nixon to accomplish similar results by executive order also failed.[11]

The Timber Supply Act was a faulty and ill-timed gambit. With complaints about harvests and harvesting methods increasing, the Forest Service and the industry were both on the defensive. To the Forest Service, the industry must join with government to reduce public fears about environmental quality. As the undersecretary of agriculture told a trade association in 1970: "To solve this dilemma requires the vision of Moses, the wisdom of Solomon, the patience of Job, and cooperation by everyone. . . . It isn't always true that the best and most economical timber practices are also the best multiple-use."[12]

It was no use. A legacy of open feuding, together with the ordinary strains of the timber-sale process, meant that there were limits to how well the Forest Service and the industry could get along. The industry kept up its criticism of the Forest Service, accusing the agency of timidity in dealing with environmental interests and of not doing enough to meet the industry's needs. The industry wanted an alliance with the Forest Service, but on its own terms. Too few of the industry's leaders were willing to admit the legitimacy of competing claims, especially for wilderness.[13]

In any event, national-forest timber became a fairly stable proportion of the national supply of wood products during the 1970s, standing at between 20 and 25 percent by 1980. The national consumption of forest products, however, fluctuated during the 1970s and into the 1980s, in particular because a series of recessions, followed by high interest rates, depressed the construction industry. It was not until the winter and spring of 1982 and 1983 that small signs of recovery developed in construction, and several lumbermills in the Northwest, closed or operating marginally for some years, resumed production.[14]

Forest products were affected importantly by inflationary price pressures during the 1970s, and it was repeatedly suggested that increases in sales of national-forest timber would reduce product prices and therefore encourage the slumping construction business. However, a congressional analysis in 1980 showed that given the proportion of supplies reflected in national-forest sales, a 20 percent increase in sales would equal only a 4.5 percent increase in total domestic production. Product prices, the study also pointed out, were more affected by processing and transportation costs than by the costs of timber. The study predicted that increases in national-forest harvests, particularly in the Northwest, would be offset by reduced harvests on private lands, given industry's reluctance to expand facilities in uncertain markets, and by reduced imports.

The study suggested that the adverse environmental effects of expanded sales would be an important countervailing factor against the benefits. A larger sales program would also require a bigger budget for the Forest Service, which would be offset to an unpredictable degree by receipts. In conclusion, the report observed: "On balance, this analysis concludes that current Forest Service timber sale policies are reasonable. An expansion has both beneficial and adverse consequences that affect different interest groups and sectors of the public. As a result, it is difficult to compare the effects."[15]

Such a thoughtful conclusion to yet another debate over the management of national-forest timber contrasted markedly with the emotional proclamations that characterized all sides of the question, especially after the early 1960s. The Forest Service had, through painful ordeal, learned to produce a program that could be described as "reasonable," but it was not the program that the Forest Service said it wanted. The congressional study discounted the Service's projections of an eternally increasing demand for forest products, because the facts of demographics, economics, and other realities could not support them.

Regarding multiple use, the report was more heartening to Forest Service philosophers than were the findings of another national body, the Public Land Law Review Commission, a decade earlier. The commission's 1970 report was relatively insensitive to environmentalist sentiments and ambivalent toward the concept of multiple use. In place of the latter, the report suggested, "management of public lands should recognize the highest and best use of particular areas of land as dominant over other authorized use." Such statements alarmed economic and environmental interests alike, and in the end, the report had little positive influence.[16]

A RARE TEST OF MULTIPLE USE

The Forest Service, despite its apparent preoccupation with timber, endeavored to make multiple use the solution to its problems during the 1970s. More was at work in the land than the sentiments that had given birth to multiple use, however. That multiple use could not answer all questions was revealed in two attempts to resolve the wilderness issue once and for all—the Roadless Area Review and Evaluation (RARE).

By the early 1970s the Forest Service was being bedeviled by uncertainties arising from wilderness proposals. Bent on increasing the annual cut from the national forests during the administration of Chief Cliff, the Service found itself repeatedly thwarted by requirements that roads and other development not be pushed into areas that were under congressional consideration for wilderness designation. Wilderness bills were introduced specifically to stop road construction, while lawsuits challenged one administrative decision after another. According to Cliff, "The offshoot of this is that it was delaying decisions on land use on the national forests, and was hampering long-range planning for both development and preservation."[17]

Cliff launched a study to examine all roadless areas and to determine finally which should or should not be wilderness, according to their qualities or the possibilities for multiple-use management. The decision to undertake a comprehensive completion of the wilderness system was not only an effort to define what ought to be wilderness; it was also an attempt to prevent wilderness from interfering with timber and development and to end the question once and for all. It did reflect a multiple-use approach to what was becoming an impossible administrative problem, however.

The work started after Cliff retired in 1972. It went fairly rapidly, producing an inventory of areas that the Forest Service believed to be qualified for wilderness, others that it thought not qualified, and some that deserved further study. That satisfied no one, of course. Proponents of timber harvesting or other development argued against every wilderness proposal, while proponents of wilderness considered all, as well as the entire program, insufficient. The latter were particularly incensed when the Forest Service seemed bent on developing areas that wilderness advocates believed deserved another look for preservation's sake.[18]

Not only were wilderness proponents unwilling to close the case on any area not proposed for wilderness designation, but many interests in the vicinity of proposed or rejected wildernesses felt themselves unfairly used. So Assistant Secretary of Agriculture Rupert Cutler and Cliff's successor, John R. McGuire, determined to try again. Deciding that the general public had not been sufficiently consulted during the first RARE, the Forest Service set out to obtain the widest general participation in wilderness decision making.

RARE II, as the renewed program was called, set off an incredible storm. Economic hard times combined with distrust of government to turn many public meetings into unbelievable tumults. Things were so bad in some places that Forest Service hearing officers required police protection. The uproar was not limited to the traditional proponents and opponents of wilderness, nor was it attributable to the immediate issues. Segments of the recreation industry fell into wrangles over whether wilderness or development would better support their interests. Hunters and off-road-vehicle enthusiasts attacked the wilderness program as a denial of their desires. Unemployed timber and lumber workers used the hearings as forums for their complaints about the national economy. Redwood loggers raised a ruckus over the Redwood National Park in California, something that was not affected by RARE II one way or the other. Positions reflecting a baffling variety of special interests were advanced, and even more unfocused displeasure was expressed. Somewhere in the uproar the Forest Service was supposed to develop a final program for national-forest wilderness.

After the steam had mostly blown off, the Forest Service did manage to emerge with a new wilderness package, once again divided among yes, no, and look again. The unique circumstances that made RARE II such a trial probably had passed. Nevertheless, it was unlikely that the issues involved in wilderness would ever disappear entirely, so long as there was one group that wanted to develop a given landscape and another that wanted it kept as is.

Although multiple use was in large part an answer to recreational and wilderness sentiment, it could not solve the entire question alone. Partly, that was because the Forest Service declined to accept wilderness as one of the multiple uses of the National Forest System as a whole. It viewed wilderness at the local level and perceived it as a place where multiple use was not allowed. Not all multiple uses were possible in a single place, because some uses—wilderness in particular—must be exclusive, or nearly so. The Forest Service did not deny the validity of wilderness; it, in fact, had invented the administrative principle and had developed a large wilderness system over several decades. But it did seem to treat wilderness as a sort of removal from the national forests, rather than one of several possible purposes of the forests. Those purposes—the five uses listed in the Multiple Use Act—were tangible and therefore would not be served by the intangible benefits that purported to emanate from wilderness. When wilderness threatened one of the material uses of a given area, the Forest Service was philosophically inclined to come down in favor of multiple use—although it was politically wise enough not to say so explicitly.

ENVIRONMENTALISM

The failure of multiple use to accommodate wilderness in a way that would satisfy everyone was not all attributable to any fine distinctions between multiple use and wilderness. Rather, the driving conservation sentiments to which multiple use had been a response—recreation and wilderness—gave way during the 1960s and 1970s to a more subtle public interest in the livability of the whole environment. That expressed itself first in a growing body of law and regulation to protect environmental quality, then in the increasing use of the courts to bring federal actions into accord with the new environmental consciousness. For the Forest Service the events meant, among others things, that the typical caseload of lawsuits against the agency grew from an average of one a year early in the 1960s to about two dozen by the mid 1970s. One of those cases nearly brought national-forest timber management to a dead stop.[19]

The United States boomed economically during the two decades after the end of World War II. National prosperity fostered enormous expansions of urban areas, with single-family houses on suburban plots epitomizing the American dream. General national wealth supported strong industrial development, with factories and smokestacks sprouting everywhere. The automobile became the quintessential symbol of well-being. The development boom came at a price, however. Urban air was befouled, and many rivers became sewers. Highway construction and urban renewal caused needless destruction and community disruption. In the 1960s, accordingly, sentiment grew for a restoration of environmental quality.

The dangers of pollution had long been well known and were the objects of some of the oldest municipal legislation in the country. Technical appreciation of the complexities of life, however, remained the province of scientists until the 1960s. When the unsightliness and unhealthiness of the national environment impressed itself upon the public, all that was required was the right catalyst to combine scientific appreciation with public concern to produce an outburst. "Ecology" became a national slogan. The old "conservation" movement became swamped in a new uprising called "environmentalism."

The catalyst was a scientist who had remarkable literary gifts. In 1962, Rachel Carson published a book called *Silent Spring*. Explaining the interaction of human conduct and the natural environment, Carson focused on the indiscriminate use of chemical pesticides. Long applauded as a means of controlling unwanted insects, they affected other things as well. Earth's creatures, including humans, said Carson, were joined in an infinitely complex and changing set of relationships. Attack an immediate problem with a simple-minded solution, and the result may be a host of other unforeseen difficulties. Boiled down to its simplest equation, Carson suggested that the thoughtless poisoning of insects would cause a silent spring, because the birds would no longer be there to sing.[20]

The impact of Rachel Carson's work was dramatic. Almost overnight the public became imbued with a sense of the complexity of the national environment, while scientists gained an increasing voice in public affairs. Acknowledging the technical difficulty of managing the environment, the nation long had left decisions to experts. That came to an end because the environmental awareness suggested that no specialist could handle all aspects of the problem. Competing specialists began to question one another, and the public at large made its own opinions known.

For land-management agencies, the first fallout of *Silent Spring* was a presidential commission on the use of pesticides, focused especially on DDT. Over the earnest protests of the Forest Service and others, in 1963 the commission's proposed restrictions on the use of chemicals were implemented.[21] There ensued two decades of dispute over pesticides, with DDT a particular source of rancor. Focusing on the protection of timber, the Forest Service continued to call for the use of DDT and other chemicals. Focusing on the environment as a whole or on public health, others objected to almost every suggested use of chemical insect and plant killers. The result has been a compromise in fact but not in principle—chemicals are still being used, but not as much as before and with much more caution. For the most part, proponents and opponents alike hold extreme and uncompromising positions.[22]

Besides restrictions on dangerous chemicals, the environmental movement had other effects on the federal government. An abundance of legislation came forth to solve particular problems, to protect endangered species, or to clean up polluted air and water. Support for such grand projects as reservoirs declined, while that for parks and recreation increased. In 1966 the Department of Transportation Act asserted that transportation projects could not come at the expense of certain values unless there were no prudent or feasible alternatives. In that same year the National Historic Preservation Act addressed historical values, establishing procedures that required consideration of them and a review of the effects upon them in any federal undertaking. The principle extended to the entire environment in the National Environmental Policy Act of 1969. NEPA, as it swiftly became known, established the Council on Environmental Quality, to which federal agencies were to submit environmental-impact statements on all "proposals for legislation of other major federal actions significantly affecting the quality of the human environment." Agencies must consider the full range of alternatives (including "no action") for all projects and submit their environmental analyses to general review and comment.[23]

The new legislation encouraged public intervention into governmental decision making. Any federal action for which legislated procedures had not been followed was subject to challenge and shutdown in the courts. Federal agencies resisted the loss of control represented by the various environmental enactments. Environmental agencies, in particular, had difficulty adjusting, as most of them were comfortable in the belief that what they did was "good" for the environment. NEPA procedures did not address good or bad, however; they merely required that all effects be considered and accounted for. One federal undertaking after another came to grief in the courts for agencies' failure to file impact statements, to assess effects, to consider alternatives, and so on. The Forest Service, the National Park Service, and other custodians of public lands were not immune.[24]

Was the conservation-oriented Forest Service prepared for environmentalism? Retired Chief Cliff admitted that it was not: "I think many people, including [myself], didn't fully recognize the snowballing effect with which this movement grew. . . . I can admit, in my own reactions I didn't fully or quickly recognize the total potential impact or the strength of the opposition that was developing against such things as clearcutting."[25]

Industrial clearcutting in patches, or "blocks," in Oregon in the 1950s. Given proper care, the Douglas-fir could regenerate in the harvested areas, but the alteration of the scenery in the interim is apparent (courtesy Forest History Society).

THE CLEARCUTTING CONTROVERSY

It was, in fact, over clearcutting that environmentalism most significantly affected Forest Service timber management. Previous controversies had focused on how much and where timber should or should not be cut. During the 1960s and 1970s, important segments of the public attacked a single harvesting technique, clearcutting, as a symbol of their more general displeasure with national-forest management. The public ultimately demanded a voice in something that the forester regarded as wholly within his technical domain—the selection of harvesting methods.

Clearcutting is a timber-harvesting method whose intent is to promote regeneration and the orderly management of tree species that are intolerant of shade. It accords with the agricultural analogy in American forestry, which treats timber as a ''crop.'' It establishes tracts of timber of the same age, more simply scheduled for future harvest. Often equated, therefore, with ''even-age management,'' it offers the attractions of convenience both in immediate harvesting and in long-term management. Clearcutting can be applied for other reasons as well, depending on local conditions or objectives. It is not as simple as it might appear, given the forester's desire for future growth. Various systems of harvest, according to conditions of species, terrain, soils, climate, and market, call for different sizes or shapes of clearcut areas and for various ways of leaving natural sources of seed. In many cases, clearcutting requires a positive follow-up to dispose of logging debris or to prepare the ground for reseeding or artificial restocking and to control brush and other growth that competes with young timber.[26]

Clearcutting was uncommon on the national forests until after World War II, except in the case of Douglas-fir. As foresters came to exert increasing influence over logging in this century, selective harvesting was favored as a corrective to the destructive effects of old industrial logging. Selective harvesting had adverse effects of its own, however, because it very often caused the removal of the best timber (''high-grading''), leaving behind a second-rate forest.

Douglas-fir was, for years, exceptional as a prominent object of clearcutting on the national forests of the Northwest. Railroad logging, leaving vast areas of clearcut, remained common through the 1920s. Later, for aesthetic and technical reasons, the size of clearcut areas was typically reduced by the Forest Service. By the mid 1930s, 40-acre patches were proposed as the general rule in Douglas-fir; this was determined by the drift of seed from surrounding sources. Shortly thereafter, researchers developed a system of ''selective timber management'' for Douglas-fir. The motivation for that was chiefly economic, allowing depression-hit lumbermen to save expenses by not harvesting entire stands. The Pacific Northwest Regional Office made selection harvesting the general rule in the Region before 1940.[27]

The results of selection—dead and downed timber—were frightful to behold, so clearcutting returned to Douglas-fir in the 1940s. In 1945 the Umpqua and Willamette national forests reinstituted clearcutting on ''relatively small areas (50 to 250 acres) of the oldest age classes, with those clear cut areas separated by areas of younger but perhaps presently merchantable timber.'' E. E. Carter found that commendable, because it reduced the danger of fire and encouraged the regeneration of Douglas-fir. Although one of his assistants expressed misgivings, Carter encouraged the extension of the practice to other forests. Regeneration on previous clearcuts had been good.[28]

Questions arose, however. In 1947 the Pacific Northwest Region laid out two large sales to be harvested in small clearcuts. The cutting areas were carefully defined in relation to logging roads and with reference to intervening patches that were to be cut later, after sixteen and thirty years, when the cutover areas should be restocked with new growth. According to Ira Mason, "A number of questions may be raised however with regard to the management prescription, such as for example, if a second cut is made in 16 years removing the remaining stand, what will be the seed source of these areas?"[29]

During the 1940s, clearcutting spread to other Regions, in none of which had it been totally unknown. In the South, where it did not become general until the 1960s, as far as Mason was concerned, "the important point is to get and keep control of cutting, unit by unit, on sufficiently small areas that the growing capacity of the soil can be used effectively." Other considerations were as important. Sportsmen around Priest Lake, Idaho, became restive about the Forest Service's harvesting plans in 1947, and their congressman proposed a wilderness to keep access roads out. The Forest Service regarded the area's timber as "the principal body of ripe and overmature timber upon which adjacent local communities and woods workers will be dependent during the next 40 or more years." It promised that "the recreation resources of the Priest Lake area will receive careful consideration when the timber access roads are constructed." Cutting would also be restricted along roads and the waterfront. The Service said that three hundred jobs for woodsmen were at stake; it did not inquire how many livelihoods depended upon recreation.[30]

The leaving of strips of timber along roadsides as a gesture to aesthetic and recreational concerns was common by the late 1940s. Ira Mason believed that the protection of roadsides could go too far, as in the Southwest, where no cutting was allowed at all on roadside strips. He said: "I am heartily in favor of permitting salvage cutting in roadside strips. Complete preservation was a sound move to get the roadside timber policy established, but prohibition against any sort of the treatment is the crudest form of control."[31]

Mason was persuaded by 1948 that clearcutting should be the rule for Douglas-fir, if the goal of reproduction for sustained yield was to be achieved. He advised the Pacific Northwest Region to eschew half measures: "There may be exceptional situations, particularly where scenic considerations are involved, where partial cutting should be used. But with present utilization possibilities there is seldom an advantage to be gained from the partial cutting system in stands of about 250 years old or older. It is delaying the time when cutting results in regeneration and new growth."[32]

Clearcutting also began to extend into other species. Studies in 1949 suggested that lodgepole pine suffered wind damage after selective cutting and could "best be managed on an even-aged basis." In the same year, however, a Forest Service scientist implied that there were serious dangers in clearcutting: "It is not simple to choose trees that grow well on bare land; also, the [ecological] balance that existed in the virgin forests was destroyed when the land was cleared." The biggest peril, he believed, was in soil erosion.[33]

Here and there, problems emerged following clearcutting, especially differing success in regard to regeneration and occasional outbreaks of insects or disease. Nevertheless, clearcutting was the preferred system in the Northwest by the mid 1950s. It appears that economic factors predominated over silvicultural ideas, particularly as foresters disagreed about alternative cutting systems. The factors that encouraged selection during the 1930s had been replaced by demands for the high production that clearcutting permitted. However, Douglas-fir clearcuts on steep, south-facing slopes or "snow flats" at high elevations simply would not regenerate from natural seeding and, in some instances, even when planted. Research began to perceive the effects of different local climates and slope-facing on reproduction, but changed priorities ended most of the studies by 1958.[34]

Douglas-fir clearcutting in the Northwest generally took the form of staggered patches in the 1950s. That had a number of technical advantages, and the seed residual timber was very subject to windthrow, but it left the forest with a strange, checkerboard appearance. Elsewhere, as in California, clearcutting increasingly was adopted to favor the reproduction of desired species over less desirable ones. In the South, however, the Crossett Experimental Forest demonstrated that the economic advantages of clearcutting might be exceeded by the silvicultural advantages of selective harvesting.[35]

Clearcutting was one amongst a bagful of silvicultural tools, and the Forest Service wanted to apply it wherever it deemed appropriate, in the interest of removing old growth, promoting new growth, and attaining maximum sustained yield. By the early 1960s, clearcutting had altered many scenic landscapes, particularly in the Northwest, and it appeared to be increasing, along with the expanding harvest. More and more recreationists were using the woods, and they were appalled by what they saw. Supporters of the Glacier Peak Wilderness in Washington alleged that the Forest Service had tried to reduce the wilderness area, then had planned to log right up to its limits. That, one of them charged, was a denial of multiple use: "In the Douglas-fir area, their basic principle appears to be that multiple use is the clear cutting of everything that can be logged. They usually clear cut and then burn over the land after they have completed their clear cutting. Since neither we nor our visitors enjoy looking at or walking through charred slash . . . this is a matter of vital concern to us.''[36]

In fact, however, some changes were under way in clearcutting in the Northwest. According to a management plan issued in 1965, there had been a steady decrease in the average size of cutting units over the past twenty years, with present policy limiting patches to about forty acres. That was in response to technical difficulties with reproduction and windthrow, not to any public sentiments; but it promised less drastic effects on the landscape.[37]

As clearcutting increased, so did public complaints. Vacationists neither knew nor cared about the distinctions between national forests and national parks. They regarded both as recreation and sightseeing attractions, and some of them objected to almost any large-scale timber harvesting that interfered with their enjoyment. Intellectually, most people appreciated the need for timber harvesting as the source for many of their goods, much as they perceived the need for prisons. But no one wanted a prison in his back yard or clearcutting on his particular pleasure grounds. An intellectual identification of the source of furniture could not overcome the emotional impact of an unfortunate spectacle that spoiled a family outing. A ranger's promise that the forest would grow back could not reassure people who were looking at stumps where they expected to see trees. Complaints about timber harvests, which focused increasingly on clearcutting, grew steadily during the 1960s. Unhappy vacationists wrote to their congressmen, and organized groups of recreationists, sportsmen, and outdoors lovers, along with tourist councils and organizations, joined in the growing ado.

The complaints could not be ignored forever, so in 1965 the Forest Service, following its accustomed tendencies, decided that a vigorous campaign of ''information and education'' would solve the problem. The Service had developed a major program to sell multiple use (and a larger budget for itself) in 1960, calling it ''A Servicewide Plan to Gear Multiple Use Management of the National Forests to the Nation's Mounting Needs.'' It dusted the idea off in 1965 and then launched a campaign under the banner of ''The Forest Service in a Changing Conservation Climate.'' Chief Cliff said: ''We need the understanding and support that comes from an informed public. [Our story] must be told and retold so that people everywhere will recognize and comprehend the forest patterns they see in America today.'' The objective was to ''demonstrate that timber harvesting under careful logging practices, including clear cutting in blocks, is compatible with other resources, makes possible new uses for the area cut, or creates opportunities for other uses not always possible in uncut stands.'' Among other things, the Service proposed to equal *Silent Spring* with "high impact publications" that "involve the reader emotionally." The "target date for significant progress" was January 1969.[38]

To suggest that bureaucrats could "involve the reader emotionally" to favor clearcutting, much as Rachel Carson's dying songbirds had made people condemn pesticides, was remarkable, to say the least. Equally noteworthy was the belief that the public would be charmed into docility by four years of "high impact publications" and other devices. People who manage trees have to be patient. It was asking a lot of an aroused public, militant environmentalists, and (especially) politicians to expect them to wait four years for something from the Forest Service besides talk.

Deciding that propaganda would end public objections to clearcutting reflected a lot about official Forest Service attitudes at that point. As it had in the controversies over sustained-yield units, it simply dismissed critics as ignorant or acting on "misinformation." The Forest Service was confident that if everyone could see things from the Forest Service's perspective, everyone would agree that what the Forest Service was doing was all right. Forestry as propaganda had always been preaching without listening. Now it became forestry as national head patting.

That strategy backfired and sorely undermined the agency's standing. A posture that relied on explanations instead of corrective action reinforced charges that the Forest Service was devoted to timber production above all else. The failure to listen to others and to appreciate that other people might have valid objectives for forest land different from timber production appeared to deny multiple use. Moreover, the Forest Service no longer had a technical monopoly on natural-resources management. Many critics of the Service's actions bore impeccable credentials and could not be dismissed lightly as being either uninformed or servants of some special interest.[39]

The propaganda campaign failed utterly, because clearcutting continued and because public sentiment against it rose with every glimpse of a field of stumps. In another attempt to defuse criticism, the Forest Service asked a group of deans of forestry schools to examine its clearcutting programs in the late 1960s. The panel returned with some technical recommendations to correct specific errors but pronounced Service policies suitable when properly observed. The deans defended clearcutting as a proper harvesting technique in appropriate circumstances. The Forest Service was satisfied with that, but many others were not.[40]

The Forest Service was forced, by the end of the 1960s, to take a hard look at the way it handled clearcutting. Its investigations acknowledged that, although some of its critics might be extreme, not all of their accusations were unfounded. Nor could management errors be excused because of demands for an expanded timber harvest. Many errors that were revealed in the self-examination were public-relations blunders—disregard for aesthetic sensibilities in clearcutting near sightseeing roads, in scenic vistas, or in tracts with unnatural, straight boundaries. Others were technical—clearcutting on inappropriate terrain, the erosion and silting up of streams, the failure to provide for regeneration. More serious were the procedural errors. Harvests had gone forward without regard to other values or economic uses that would be disrupted by clearcutting instead of selective cutting. Instances in which harvests that were supposed to attract or support minor local lumber industries had severely disrupted more important recreational and sporting industries were especially questionable if the purpose of the national forests was to serve community needs—and if multiple use was to be their governing policy.[41]

The Service's self-examinations were mostly technical; they did not often acknowledge that it might be advisable to reserve some timberlands for other uses. Critics were not satisfied that timber would not always come first in the Forest Service's interpretation of multiple use. That feeling was exacerbated by the fact that by 1970 the Forest Service's position had become one of defending the legitimacy of clearcutting as a management practice, instead of acknowledging that there might be cases when the "best" management practice from the forester's standpoint might not be the best from other perspectives.

That narrow technical defense of clearcutting itself was undermined by the views of experts not fighting for a bureaucratic position. The 1970 edition of the standard textbook on silviculture said the practice was often based on outdated and incorrect assumptions, concluding that "it is actually fortunate" that clearcutting "has worked at all."[42]

More explosive was a report on Forest Service timber management in Montana, prepared by that state's School of Forestry. Dean Arnold W. Bolle assembled his own experts at the request of Senator Lee Metcalf, who had been bombarded with complaints about what the Service was doing. Bolle's panel discovered that even from the narrowest silvicultural standpoint, several large harvests were entirely inappropriate, ruinous to future forest growth, and thoroughly destructive of values that were more important than timber. In clearcutting where it should not and in ignoring everything but timber-sales targets, said the experts, the Forest Service was actually practicing timber mining. "Multiple use management, in fact, does not exist as the governing principle," they concluded.[43]

Bolle presented his report to Metcalf in November 1970, and the senator promptly released it to the public. The Forest Service was devastated, partly because Bolle's report upstaged the Service's own self-examinations but mostly because the document made the Service's protestations about multiple use seem empty and its claims of professional correctitude false. No longer could the Forest Service presume to speak for all of forestry or to say that its actions reflected universal technical judgments, unalloyed with other considerations. Multiple use appeared to be a meaningless slogan, hiding a devotion to timber production.

The Forest Service made some serious mistakes in the controversy over clearcutting, and it paid a severe price for them. Edward C. Crafts, a former Forest Service man, publicly described the agency in the early 1970s as demoralized. Cliff himself asserted: "Many employees have recently expressed concern on the direction in which the Forest Service seems to be heading. I share this concern. Our programs are out of balance."[44]

Cliff took steps to correct the imbalances, which he blamed on the budgetary process. Congress and the several presidential administrations were considerably more willing to support programs that produced revenues than those that did not, at least in the short run. To that sentiment was added the political impact of ceaseless calls for more industrial access to the national forests; in the early 1970s that turned on beliefs in a national housing shortage. Cliff and his successor capitalized on the clearcutting controversy to produce more funding for other programs, although never as much as they said they wanted.[45]

Cliff, without admitting that anything had previously been wrong, instituted more careful controls on harvesting practices, including increased sensitivity to aesthetics. Landscape architects helped to design clearcuts so that they would assume a more natural appearance than the square edges and corners that had been common earlier. Modifications in clearcutting also followed changing silvicultural notions, particularly in the Northwest. By the 1970s the Forest Service had turned increasingly to "shelterwood" harvesting (taking about half the timber in a stand) in the Northwest, at least on arid sites.[46]

The Service also supplemented its customary propaganda with attempts to involve the public in decision making. Eventually, some rather elaborate programs with names like "Inform and Involve," advisory committees, and a reformed planning process elicited public sentiment before final decisions were made. Requirements of the wilderness program and of NEPA further encouraged public participation. Nevertheless, many people remained convinced that the agency was not really listening but was merely conducting an empty exercise in public relations.

The clearcutting controversy lumbered toward its big showdown in 1971. That year, the Forest Service issued a booklet called *National Forest Management in a Quality Environment: Timber Productivity.* It listed thirty "problems" that the Service wanted to correct in its various programs, some of which it could resolve alone, more of which involved bigger budgets. Half of the booklet was given over to an explanation of timber management and its purposes, including descriptions of harvesting systems. Clearcutting received disproportionate explanation and justification, with several attractive photographs of harvested areas. Cliff called the booklet "a major step forward in our continuing efforts to achieve two of our major objectives—timber production and environmental protection and enhancement." He promised to adopt the changes in policies and practices required by the thirty problems and suggested that the booklet would "be used in connection with the hearings that Senator Church has scheduled."[47]

Senator Frank Church held "informational" hearings on the clearcutting issue in the spring of 1971. Forest Service officials, including Cliff, testified, as did their outstanding critics and a bevy of foresters, scientists, environmentalists, industrialists, and others of varying persuasions. Extremist cant was heard on both sides, but mostly the committee received a number of thoughtful presentations on all sides of the many questions involved. The panel decided that no changes in law were necessary and that clearcutting was an important management tool when used appropriately. Satisfied that the Forest Service had moved to correct previous errors, the committee did not support a ban on clearcutting.[48]

In the normal course of public uproars, that should have ended the clearcutting controversy. The Forest Service had tightened up its procedures, had become more circumspect about clearcutting, and now provoked less complaint (while in old clearcuts, new growth rose to hide the scars). There remained only the usual spate of publications from diehards on both sides who wanted either more clearcutting or none at all. There was a brief flurry of attention in 1972, when rumors circulated that President Nixon had considered banning clearcutting by executive order but had caved in to industry opposition. The press, however, tired of the issue after the Church Committee's hearings.[49]

A PEACEFUL INTERLUDE

Chief Cliff retired in the aftermath of the Church Committee's hearings. He recommended only one person as his successor—John R. McGuire, who had been Cliff's representative in political and interagency negotiations during the late 1960s and early 1970s. As Cliff put it, "The thing that impressed me most about him, and the thing that I felt was badly needed, was the ability to negotiate with other people and to do it in a very nice way."[50] McGuire would need his persuasive abilities, for the clearcutting controversy would return to vex the Forest Service one last time. That was not apparent when he became chief in 1972, however. His first objective was to find a way out of the ceaseless conflict, confrontation, and frustration associated with endless controversies over wilderness, clearcutting, and whatnot.

McGuire's first major triumph was the Forest and Rangeland Renewable Resources Planning Act of 1974 (usually shortened to RPA). It was an attempt to adjust the appropriations processes that left the Forest Service budget unbalanced and to avoid the conflicts generated when any single program was strengthened. RPA directed the secretary of agriculture periodically to assess the national state of forest and rangeland resources and regularly to submit recommendations for long-range Forest Service programs to meet future needs. Recommendations were to cover all activities of the Forest Service. The act attempted to regularize the budgetary system by requiring the president to state his policy on Forest Service budgets over periods of five and ten years and to explain the relation of each year's budget to the long-range plan. The act also required interdisciplinary planning and the maintenance of a current inventory of national-forest lands and resources. All necessary work on the national forests was intended to be up to date, with backlogs eliminated, by the year 2000.[51]

RPA was passed just after the Forest Service issued its "Environmental Program for the Future." Taken together, the program and RPA were represented as encouraging the intensive but coordinated management of all resource values so as to avoid the disputes that had characterized timber harvesting and wilderness designations in the immediate past.

In actuality, RPA was an answer to a bureaucrat's prayer. It superficially realized longstanding visions in the Forest Service of a comprehensive national forestry plan based on a comprehensive inventory, with the Service producing both according to its own judgment. More practically, RPA authorized the endless generation of paperwork, which necessitated the expansion of the agency's work force. Politically, the production of programs effectively defused arguments that the Forest Service concentrated too much on timber and not enough on other things. The Service now assembled, in the name of multiple use, impossibly expensive plans to manage everything at a high level. When Congress did not fund the entire program as proposed but chose to keep the total Service budget within reason, it was not the Forest Service's fault. Any failure to realize multiple use, any emphasis on timber over other programs, could be attributed to Congress's refusal to fund all programs as proposed.

RPA, in other words, was the Forest Service's shield against the slings and arrows of outrageous critics. It was not enough, however. As the RARE exercises demonstrated, the Forest Service could not avoid all conflict. The last round of the clearcutting furor was about to open. This time, bitter public debate was replaced by the firm, indeed hard, words of a judge.

THE MONONGAHELA SHUTDOWN

It started with some turkey hunters, before McGuire became chief. As Cliff recalled the first, unrecognized warning, a small delegation visited the Department of Agriculture early in the Nixon Administration to protest clearcutting on the Monongahela National Forest in West Virginia, because they feared the destruction of turkey habitat in areas they had used for generations. Cliff was sympathetic, ''But my initial reaction was—what's more important, the personal pleasure of a small handful of people for turkey hunting, or the utilization of this resource for production of jobs and raw material, done under a silvicultural system that, if properly applied, was sound?'' Cliff was not about to apologize for not having recognized as legitimate another view of the best use of a resource but merely for having failed to gauge the power the competition could bring to bear. ''What I didn't realize is how potent they could be in expanding this protest,'' he said later.[52]

Potent they were indeed. After repeatedly trying to get the Forest Service to recognize their concerns, the hunters and their friends went to court in 1973. The Izaak Walton League soon joined in their action, which hit the Forest Service in an unexpected place. The plaintiffs contended that in embarking upon a clearcut, the Service violated the letter and the clear intent of the Organic Act of 1897, which first authorized timber harvests on the national forests. The act was the only real charter that the timber program had.

The 1897 legislation said: ''For the purpose of preserving the living and growing timber and promoting the younger growth on national forests, the Secretary of Agriculture . . . may cause to be designated and appraised so much of the dead, matured, or large growth of trees found upon such national forests as may be compatible with the utilization of the forests thereon'' and ''such timber, before being sold, shall be marked and designated.''[53] The Forest Service's *Manual,* as amended in 1957, had something different to say: ''Maturity, as used in the Act, means a condition which makes it desirable to cut the tree . . . ; when a tree should be removed so as to permit increased development of the remaining stand; and when a tree has reached its highest value for some product.''[54]

The lawsuit maintained that in cutting all timber, young as well as mature, the Forest Service exceeded its authority. It also violated the marking requirement, because it did not mark all timber in a clearcut before sale. When the government argued that changing silvicultural requirements and national demands for timber required a free definition of *matured* as used in the 1897 legislation, the judge referred to the dictionary and denied the argument. If the law governing timber sales required changes for modern times, said the court, the law should be changed by Congress, not by the bureaucracy. An injunction stopped all timber harvests on the Monongahela except those which were strictly in accord with the law; in other words, clearcutting was forbidden. An appeals decision in 1975 made the restriction binding throughout the Fourth Circuit.[55]

Almost every flaw in Forest Service timber management came to the fore in the events leading to the Monongahela shutdown. To begin with, clearcutting in eastern mixed hardwoods was a new practice for the Forest Service, which derived it from experiments during the 1950s and 1960s. That made it dramatically new for the Monongahela area when it was introduced. There, the Service had spent decades decrying the ''destructive'' clearcut harvests that had preceded its acquisition of the land. The local foresters were content with the rightness of their decision. Not only did they fail to prepare the public for a change in procedures, but evidently they made little effort to explain the rationale to the Service's own working-level employees, who told their neighbors that the Forest Service was about to conduct ''destructive'' harvesting.[56]

The Forest Service's easy dismissal of objections, as in the sustained-yield controversies of the 1940s and the clearcutting fuss of the 1960s, remained habitual during the 1970s. Sportsmen who complained about the Monongahela harvest were regarded as selfish or misinformed and their grievances as, perforce, baseless. That attitude was bound to arouse resentment, particularly in West Virginia. In much of Appalachia, hunting was no idle pastime; it was a traditional and essential ritual of manhood, and for many families in that poor region, it provided meat that was otherwise unavailable. Anything that threatened hunting seriously threatened that segment of the population, culturally, psychologically, and economically. The Forest Service appeared downright callous.

The Service also misled itself with some of its own long-held and unquestioned assumptions. The Monongahela management plan offered customary platitudes about clearcutting's being good for wildlife. Wildlife meant deer to the foresters; when the observation was first made, it was drawn on clearcuttings in the West, where deer constituted an important hunting resource. In West Virginia, hunting was mostly for rabbit, squirrel, and turkey, all of which could be seriously disrupted by clearcutting. Insofar as the original protesters were concerned, what was really at issue was, not harvesting in general or clearcutting in particular, but stands of nut-bearing old-growth trees that had survived earlier industrial harvesting. They supported abundant populations of squirrel and turkey; if the old-growth went, so would the game. A compromise that protected the vestiges of old growth might have avoided the showdown in court, but the Forest Service was not in the mood for compromise.

REPAIRING THE DAMAGE

The government decided that further appeals would be fruitless after the Circuit Court upheld the Monongahela decision in 1975. Other lawsuits promised to extend the strict interpretation of the 1897 legislation around the country, reducing national-forest timber harvests by at least half. Clearly, as the judge in the Monongahela case had suggested, a legislative solution was needed.

It arrived with the National Forest Management Act (NFMA) of 1976. That legislation was the result of a compromise between forces that supported the widest possible latitude in forestry practices and others that desired highly prescriptive legislation for the national forests. The act emphasized land-management planning, timber-management actions, and public participation in decision making. It set forth broad policies for the national forests, including some specifics that had formerly been discretionary, but it allowed the Forest Service freedom to apply its own professional judgment to forest management. That included harvesting; clearcutting was not outlawed.[57]

While NFMA offered a fundamental charter for the Forest Service, it was a refined endorsement of multiple use. The options that it preserved for the land manager ensured that the Service, and particularly its timber management, would never be relieved of controversy. Among early sources of contention under the law were "land not suitable for timber production," "optimum" and "appropriate" methods of timber harvest, "overall multiple-use objectives" for land-management plans, determinations of the "culmination of mean annual increment," and the authority to depart from the "quantity which can be removed from the forest annually on a sustained-yield basis" or "nondeclining even flow.'' Before the ink had dried on the act, controversy arose over sealed versus oral bidding for timber sales.

Not so obvious in the words of the legislation was the way it built upon the bureaucratic self-protection that the Forest Service had begun fashioning under RPA. NFMA was an ironic conclusion to the clearcutting controversy, because it diverted all arguments from the real issues in the woods into the mires of planning. The act launched an enormously detailed and complex planning process which involved the eternal generation of turgid documents to be reviewed and revised forever. Planning became an end in itself, and in the years after 1976 the Forest Service wrapped itself in red tape. Arguments over clearcutting and the like became arguments over drafts of plans, which were challenged more often on their economic analyses than on what they meant in terms of timber or scenery or turkey habitat. The Service went on its way, its fundamental outlook and approach to the forests much as they had always been.

NFMA essentially ratified what the Forest Service was doing at the time. The Service had never abandoned the old-time religion of more-wood-for-the-nation that had blessed its birth. When the opportunity presented itself in the 1950s, the Service answered the call of maximum sustained yield. There followed decades of hard knocks, as the bureau resisted the fact that wood was not all the nation needed or wanted from the national forests. Controversies, lawsuits, and new laws all competed for the Service's attention, and the Service responded with propaganda and appearances. By the mid 1970s the Forest Service believed it no longer appeared as single-minded as when its ''high impact'' publications promoted the virtues of timber harvesting.

Appearance was not substance, however. Talk of balanced programs, interdisciplinary planning, and multiple use could not alter the Forest Service's focus on timber. It was, after all, a federal bureaucracy, and accordingly it was affected by the flow of power in the government. By its nature as a revenue-producing activity, timber management figured in national economic calculations. Congress and successive presidents reinforced the Service's timber orientation by calling for expanded national-forest sales to support economic growth or to provide federal revenues. The latter became especially important during the administration of Ronald Reagan, whose materialistic attitude toward natural resources exceeded that of the federal foresters. Sales and harvests, however, never increased as much as all the presidents wanted. The Service did not receive the full budgets it claimed were essential in order to make greater sales. In the spirit of RPA the politicians, not the Forest Service, could be accused of not practicing multiple use.

Whatever the politics, the Forest Service was not driven toward timber production against its will. On the contrary, the agency was still run by foresters, and as Paul Herbert pointed out in 1953, foresters went to school to learn how to harvest timber. That had changed by the 1970s, after the forestry schools broadened their curriculums to emphasize the management of forest ecosystems as a whole. But the foresters who were running the Forest Service had been out of school for a long time, and they still seemed to be devoted to timber production. It was repeatedly suggested in public that as long as foresters ruled the Forest Service, timber would always come first among the multiple uses. It remained for the foresters to demonstrate that they could move over and grant others their full role in a land-management system that should be interdisciplinary.

Chief McGuire left the Forest Service in 1979, his agency enjoying relative peace after so many years of turmoil. His replacement was R. Max Peterson, an engineer who had a Masters degree in public administration and a varied career in the Forest Service behind him. That should satisfy the critics. No one would accuse an engineer of being obsessed with timber (engineers hear the same complaints about dams and highways). Furthermore, a background in public administration suggested preparation for multiple use, because a professional administrator should display no bias toward one resource or another. Timber management would remain important, but as far as the Forest Service was concerned, the new legislation and Peterson's appointment proved that timber was one among equals in the multiple use of the National Forest System.

The Forest Service's standards of proof on such matters continued to be more lenient than those of the public. The agency had no difficulty in persuading itself that its pronouncements and proclaimed new image reflected its real soul. All it really wanted was the trust that it had enjoyed in the salad days of the Progressive Era but had somehow lost since then. It wanted the people to trust the rightness of the Service's motives, the correctness of its ways, and its assurances that it really had taken multiple use to heart. Once again, it expected to receive trust for what it said, rather than for what it did, because the Forest Service did not believe the people could understand the complexities of its technical work. By the 1980s, however, the people had become skeptical of all governmental pronouncements. They agreed with Mr. Dooley. "Thrust ivvrybody," he advised in 1900, "but cut th' ca-ards." The Forest Service's timber-management program was no longer the exclusive property of the Forest Service. The owners of the national forests were on the alert, and they had more at stake than wood.

EPILOGUE

A fanatic is a man that does what he thinks th' Lord wud do if He knew th' facts iv th' case.

—Mr. Dooley (Finley Peter Dunne, 1900)

During the controversies of the 1960s and 1970s, some people insisted that the Forest Service had changed, that it was no longer the same dedicated agency that formerly had guarded the forests. They were wrong. The Forest Service was very much the same, and the very nature of the controversies revealed how consistent it had been in its attitudes and actions for decades.

Gifford Pinchot was wont to remark, "I have been a governor now and then, but a forester all of the time." As with the father, so with the sons. People who erroneously call Pinchot's agency the "forestry" service make an unintentional but reasonable assessment.

The Forest Service assumed charge of the national forests in 1905, when forestry represented the current wisdom on forest management. Only foresters knew all that there was to know about forests. Timber was their totem, a touchstone of civilization that was believed to be in dire peril. The timber famine threatened, and only forestry could avert it.

The Forest Service proclaimed its great mission from the beginning. It safeguarded the country's insurance against timber shortages; when conditions proved just right, the national forests would produce wood for the nation. The opportunity presented itself after World War II, and the Service seized the moment. Expanding timber harvests in the 1950s and 1960s were new only in scale. They aimed at the conversion of the forests to the condition of maximum sustained yield, which had been the Forest Service's objective from the outset. Big harvests surprised the public; the controversy they engendered surprised the Forest Service.

The Forest Service took pride in the fact that its history included a succession of great controversies. Those were often of its own making—natural expressions of the Service's special mission. For the most part, the disputes had been with the timber industry over regulation. During the Progressive and New Deal periods, when industry was a national punching bag, it was easy to identify the opposition as being in the wrong, the Forest Service as being always in the right. In the late 1940s, when other interests challenged the Service's authority over grazing, a national chorus supported the Forest Service against the evildoers. Combined with the sense of mission, the legacy of controversies made the Forest Service assume that its actions should not be questioned because its motives were honorable. The organization also adopted the belief that any opponent was perforce in the wrong, whether from ignorance or from evil intent. Ignorance could be corrected with "information and education"; evil simply resisted. When the Forest Service believed the time had come to fulfill its mission, the stage was set for a new kind of controversy.

The Forest Service did not alter its outlook, and that was the source of its difficulties in the 1960s and 1970s. Focused on the goal of maximum sustained yield, it did not confront changes in the national culture. Forestry no longer represented the collective national wisdom on forests. It was not the way to achieve everything that the public now wanted from its national forests, a half-century or more after they were created.

The Forest Service appeared to accommodate the national climate when it embraced multiple use and when it hired landscape architects, wildlife biologists, and other nonforesters to work in its programs. That was not a real adaptation to the new order, however, as the new specialists were there largely to refine commodity production. Multiple use expressed the materialism that underlay the agency's foundations. It therefore could not accommodate nonmaterial interests in such things as wilderness, scenery, and nongame wildlife—particularly when the nonmaterial called for lesser production of the material.

The Forest Service was also bent for conflict by adhering to another outlook that characterized its origins. In all aspects of American life the Progressive Era was the heyday of the expert, the professional technician, the technocrat. National faith in technocracy faded over the following generations, but the Forest Service remained technocratic in outlook, persuaded that only it could make the correct decisions for the national forests. When outsiders challenged the Service, they challenged its professional ego.

Feeling supreme in its own domain, the Forest Service did not question its devotion to timber production. Criticism from within was unlikely, because the shared culture of the agency was pervasive. Criticism from without was not to be acknowledged as legitimate; two generations of controversy had sustained the Forest Service's rectitude. When opposition arose against sustained-yield units, it had to be dismissed, lest the Forest Service admit error. When it arose against clearcutting, it had to be dismissed again, because it questioned the forester's professional judgment.

The nation surrounding the Forest Service was considerably different in the 1960s from what it had been in 1905. Now botany, ecology, zoology, and a host of other disciplines challenged forestry's monopoly. The other fields had different objectives for the forests. The general public no longer perceived the national forests merely as storehouses of timber, but instead as recreational wonderlands, threatened by sustained-yield goals. Now only the timber industry championed the Forest Service's focus on timber.

The agency had to change. That did not happen easily or without pain. It required vicious controversy, restrictive legislation, lawsuits—and a growing volume of criticism from the forestry community at large, which adapted to new realities far more easily than did the Forest Service.

The Forest Service now says it regards timber as but one of several multiple uses of the forests. It still makes extravagant predictions of the future national need for timber, and it has not abandoned in principle the goal of maximum continuing productivity; but it does profess the legitimacy of competing values. Albeit with some difficulty, the Forest Service now says that certain nonmaterial purposes might validly be served by giving up the material benefits of some timber. In saying such things, however, the Forest Service calls them "concessions," implying that it has not yet embraced multiple use in all its ramifications. The listener is left with the feeling that beneath the surface, timber is still first among multiples.

Reflecting the Forest Service's new public image is the latest incarnation of sustained yield—"nondeclining even flow." It took decades to settle on a definition of sustained yield. To the earliest foresters it meant maintaining production by replacing harvest with new growth—cut one, plant one, in its most simplistic form. Increasing sophistication taught that respect for nature's own forces could accomplish the same goal. Appropriate harvesting and cultural practices could ensure that new growth would replace what was taken away. That eventually was carried to its logical conclusion of replacing old volume with new volume—meaning the conversion of static, old-growth forests to fast-growing new forests. It became theoretically possible to consider a maximum rate of sustained yield through complete conversion, reproducing and later harvesting the greatest possible volume of timber.

Realization of maximum sustained yield never did prove possible, for various reasons. First, industrialists transformed it into a more economical concept—sustained yield meant managing raw materials so as to sustain the industries and the communities that depended upon them. That did not necessarily mean converting the nation's forests to maximum-volume production. Later, political considerations intervened—people besides foresters and industrialists perceived values in old-growth forests other than timber and wanted them left as they were.

The increasing harvests of the 1950s represented a temporary conjunction of industrial and Forest Service interests. Industry wanted the national forests' old-growth timber because it was the greatest remaining part of that resource; when that became scarce, industry would find other sources or substitutes. To the Forest Service, however, the demands of the 1950s provided an opportunity to convert the old growth to second growth (and vindicated its predictions of eternally increasing demand). By 1960, sustained yield meant, according to law, "the achievement and maintenance in perpetuity of a high-level annual or regular periodic output" of timber "without impairment of the productivity of the forest."

That proved to be impossible. Predicted increases in demand did not develop after the mid 1960s, by which time other considerations—budget, politics, competing uses and philosophies—had also retarded the approach to maximum sustained yield on the national forests. Sustained yield entered the period of multiple use and became difficult to support. Concentrated attention to one resource compromised other values that multiple use was supposed to accommodate. By the time the National Forest Management Act appeared as the ultimate triumph of multiple use, the goal of maximum sustained yield was futile. The rigid materialism of sustained yield was not flexible enough. Something more adaptable was in order, and this was expressed as nondeclining even flow—a longer-viewed, nationwide approach to the timber resources of the national forests.

The new form of sustained yield bears some resemblance to its predecessors, for at heart it means a continuous production of forest products. However, it does not rigidly measure production annually on each parcel of ground. It seeks, instead, a regular contribution to the national timber economy from the National Forest System as a whole—a little more here, less there, possibly varying from year to year, so long as the contribution overall is continuous. That is, resources should be extracted in such a way that general productivity will not decline.

Exactly what (if anything) nondeclining even flow will mean and how it will work must be hammered out in practice. It may turn into vicious trades—logging here to compensate for wilderness there; or it may conjoin the two oldest concepts of sustained yield—the forester's interest in sustaining the forest and the industrialist's in sustaining the forest user—thus ending decades of controversy. Whether it will at long last integrate timber management into multiple use, however, remains to be seen.

The Forest Service remains a forestry service at heart. Its collective outlook is still that of forestry, with all of its traditional attention to timber. That could change (but will not completely disappear) in coming years. Foresters now make up only a part of the Service. The law now requires that national-forest management be interdisciplinary, and perhaps eventually it will be. Moreover, forestry education has changed; it now emphasizes "multiresource" forestry over the timber primacy of the past. Foresters of the future should be more broad-minded than many of their predecessors. The wood chopper's voice will remain important, but some day it just might cease to be the dominant one in the Forest Service.

The history of Forest Service timber management is only reflected in the events of the last two decades; it is not encompassed in them. Whether measured by forestry's own original goals or by the national sentiments of the disputatious present, timber management on the National Forest System has been a success. Mistakes have been made, to be sure, and principles and techniques have changed; but that was inevitable. The national forests survived a depression and two world wars, and when the nation needed their timber, it was available. Today they provide about a quarter of the nation's supply of raw forest products. That is not as high a proportion as the Forest Service predicted or wanted, but it is significant. The fact that most of the rest of the nation's timber supply comes from private lands is a credit to the influence and example of the Forest Service. The timber famine never developed.

The Forest Service has not always met everyone's approval, but the Forest System survives in its hands. For a group of mere mortals who for a long time believed themselves to be burdened with a sacred mission, perhaps the Forest Service has done well enough by its office on earth.

NOTES

Record numbers for items in the regional offices of the National Archives and Records Service are abbreviated thus: accession number 54-A-111, record number 59858, is listed as 54-A-111/59858. Records in the National Archives, Washington, D.C., Records of the Forest Service, Record Group 59, are identified by record entry, followed by RG59/the box number. The C. J. Buck Collection, Timber Management Office History File, Pacific Southwest Regional Office, Forest Service, San Francisco, is cited as R5-Buck. These acronyms are used in citations throughout the notes:

D	Denver, Colorado
DF	District Forester
DTM	Division of Timber Management
DTMRF	Division of Timber Management Reading File
F	*Forester*
FH	*Forest History*
FHS	Forest History Society
FQ	*Forestry Quarterly*
FS	Forest Service
FSYU	Federal Sustained Yield Unit
FW	Fort Worth, Texas
GPO	Government Printing Office
JF	*Journal of Forestry*
JFH	*Journal of Forest History*
LN	Laguna Niguel, California
NA	National Archives
NARS	National Archives and Records Services
NF	National Forest
NLMA	National Lumber Manufacturers Association
NPS	National Park Service
PIO	Public Information Office
PNWRO	Pacific Northwest Regional Office
RF	Regional Forester
RG	Record Group
S	Seattle, Washington
SB	San Bruno, California
SWRO	Southwest Regional Office
SYU	Sustained Yield Unit
TM	Timber Management
TMO	Timber Management Office
TSO	Timber Sales Office
USDA	United States Department of Agriculture
USFS	United States Forest Service

PROLOGUE

1. Quoted in Frederick Jackson Turner, "The Significance of the Frontier in American History" (1893), in his *The Frontier in American History* (New York: Henry Holt, 1920, 1947), 14.

2. There are several good surveys of the origins of sentiment for forest conservation in America. The following discussion relies especially on Harold K. Steen, *The U.S. Forest Service: A History* (Seattle: University of Washington Press, 1976), 3–46, and on the excellent chronological documentation in Norman Wengert, A. A. Dyer, and Henry A. Deutsch, *The "Purposes" of the National Forests—A Historical Re-interpretation of Policy Development,* a report prepared under contract between Colorado State University and the Forest Service (Fort Collins: Colorado State University, 1979).

3. For a good account of the relationship between fuel availability and the shift to stoves see A. William Hoglund, "Forest Conservation and Stove Inventors, 1789–1850," *FH* 5 (Winter 1962): 2–8. Hoglund says that the general adoption of stoves during this period actually set the conservation movement back by several decades, because their efficiency relieved immediate fears about the depletion of forest resources. Evidently, the army was the first federal agency to show concern about the progressive scarcity of fuel wood (as the navy had previously regarding shortages of ship timbers); as early as 1844 the quartermaster general predicted increasing fuel prices as deforestation proceeded (see David A. Clary, *These Relics of Barbarism: A History of Furniture in Barracks, Hospitals, and Guardhouses of the United States Army, 1800–1880,* Report DAC-7, prepared under contract for the NPS (Bloomington, Ind.: David A. Clary & Associates, 1982), 48–49; republished as *These Relics of Barbarism: A History of Furniture in Barracks and Guardhouses of the United States Army, 1800–1880* (Harpers Ferry, W. Va.: NPS, 1985), see p. 70.

4. George Perkins Marsh, *Man and Nature* (1864), ed. and with an introduction by David Lowenthal (Cambridge: Harvard University Press, 1965).

5. Marsh, *Man and Nature,* 204.

6. A board foot is the standard unit of volumetric measure for sawtimber. It is normally rendered as BF (board feet), MBF (one thousand board feet), or MMBF (one million—a thousand thousand—board feet). A board foot equals the amount of lumber in a slab one-inch thick and one-foot square. A 16-foot log, for example, with the small end measuring 21 inches in diameter, includes 300 BF (Scribner Decimal C Scale). F. C. Ford-Robinson, *Terminology of Forest Science, Technology, Practice, and Products* (Washington, D.C.: Society of American Foresters, 1971), 27. The following are typical estimates of supply and predicted duration of timber in the United States, 1877–1919:

Year	Estimated Supply (in billions of BF)	Source of Estimate	Duration (in years)
1877		Secretary of Interior	20
1880	856	Tenth Census	
1896	2,300	USDA Division of Forestry	58
1898	1,400	G. W. Hotchkiss	
1900	1,390	Twelfth Census	
1902	2,000	USDA Division of Forestry	
1905	1,970	*American Lumberman*	66
1907	2,500	USDA Division of Forestry	71
1911	2,800	Department of Commerce	55
1919	2,500	Society of American Foresters	50

Sources: Annual Report of the Secretary of the Interior, 1877 (Washington, D.C.: GPO, 1878); USDA, Division of Forestry, *Facts and Figures Regarding Our Forest Resources Briefly Stated,* circ. no. 11 (Washington, D.C.: GPO, 1896); R. S. Kellogg, *The Timber Supply of the United States in 1905,* FS circ. no. 52 (Washington, D.C.: USDA, FS, 1906); R. S. Kellogg, *The Timber Supply of the United States,* FS circ. no. 97 (Washington, D.C.: USDA, FS, 1907); U.S., Department of Commerce, Bureau of Corporations, *Report on the Lumber Industry,* 4 pts. (Washington, D.C.: GPO, 1913–14); Samuel Trask Dana, *Forest and Range Policy: Its Development in the United States* (New York: McGraw-Hill, 1956), 78, 192; Jenks Cameron, *The Development of Governmental Forest Control in the United States* (Baltimore, Md.: Johns Hopkins University Press, 1928), 285; and Society of American Foresters, *Forest Devastation: A National Danger and a Plan to Meet It,* reprinted from *Journal of Forestry,* Dec. 1919.

7. 26 Stat. 1103.

8. 30 Stat. 34, 35, 36, as amended. This part of the Sundry Civil Expenses Appropriation Act for Fiscal Year 1898 is regarded by the Forest Service as the "Organic Administration Act of 1897," hereafter cited as "Organic Act."

CHAPTER ONE. THE NATIONAL FORESTS AND THE STRUGGLE FOR CONSERVATION

1. U.S., Bureau of the Census, *Statistical History of the United States from Colonial Times to the Present,* Historical Statistics of the United States, Colonial Times to 1970, prepared by the United States Bureau of the Census (New York: Basic Books, 1976), 533; Steen, *U.S. Forest Service,* 26–37, 74; Richard C. Davis, ed., *Encyclopedia of American Forest and Conservation History,* 2 vols. (New York: Macmillan, 1983), 2:820. For one recent study of the role played by watershed protection in forest reservation see Ronald F. Lockmann, *Guarding the Forests of Southern California: Evolving Attitudes toward Conservation of Watershed, Woodlands, and Wilderness* (Glendale, Calif.: Arthur H. Clark, 1981).

2. Dozens of documents quoted by Wengert, Dyer, and Deutsch, in *"Purposes" of the National Forests,* hold up European forest stewardship as the example that Americans should follow. The European evocation was continuous from *Man and Nature* (1864) to the 1897 legislation.

3. The intricate political history of these events is beyond the scope of this study; it is covered very well by Steen in *U.S. Forest Service,* 47–75. The Interior Department soon had cause to regret the excision of the reserves from its jurisdiction over the public domain, and within a few years a tradition of bureaucratic bad blood between the Forest Service and Interior was firmly established.

4. Filibert Roth, "A Word to the Members," *JF* 16 (Feb. 1918): 145; James W. Toumey, in *Biographical Record of the Graduates and Former Students of the Yale Forest School* (New Haven, Conn.: Yale Forest School, 1913), 10 (for "propaganda"; cited hereafter as Yale, *Biographical Record*); (Bernhard E. Fernow), *FQ* 1 (Oct. 1902): 1 (for "no past masters").

5. Ralph S. Hosmer, with Bruce C. Harding, "Early Days in Forest School and Forest Service," *FH* 16 (Oct. 1972): 8 (based on quotation of German texts); "Sir William Schlich," and W. B. Greeley, "A Memorial Fund for Sir William Schlich," *JF* 25 (Jan. 1927): 5, 10. Schlich's work was published as *Schlich's Manual of Forestry* in several editions by Brandbury, Agnew & Co., London. It was in its fourth edition in 1910.

6. Henry Solon Graves, *The Principles of Handling Woodlands* (New York: John Wiley & Sons, 1911), vi; Henry Solon Graves, *Forest Mensuration* (New York: John Wiley & Sons, 1913), v. Graves copyrighted the latter text in 1906.

7. A. B. Recknagel, *The Theory and Practice of Working Plans* (New York: John Wiley & Sons, 1913), iii; Herman Haupt Chapman, *Forest Valuation* (New York: John Wiley & Sons, 1914), v; James W. Toumey, *Seeding and Planting* (New York: John Wiley & Sons, 1916), v.

8. Gifford Pinchot, *Breaking New Ground* (New York: Harcourt, Brace, 1947), 30, 152, and "The Profession of Forestry," *F* 5 (July 1899): 157; the Pinchot family declaration is quoted by Susan L. Flader in "Thinking Like a Mountain: A Biographical Study of Aldo Leopold," *FH* 17 (Apr. 1973): 17.

9. Carl A. Schenck, "The Training of Professional Foresters in America," *F* 5 (May 1899): 105–6.

10. Bernhard E. Fernow, *Economics of Forestry: A Reference Book for Students of Political Economy and Professional and Lay Students of Forestry* (New York: Thomas Y. Crowell, 1902), 137, 171–72; Henry S. Graves, *Practical Forestry in the Adirondacks,* USDA, Division of Forestry Bulletin no. 26 (Washington, D.C.: GPO, 1899), 12–13.

11. M. Nelson McGeary, *Gifford Pinchot: Forester-Politician* (Princeton, N.J.: Princeton University Press, 1960), 19; Pinchot, *Breaking New Ground,* 106–7.

12. Andrew Denny Rogers III, *Bernhard Eduard Fernow: A Story of North American Forestry* (Princeton, N.J.: Princeton University Press, 1951); see also Steen, *U.S. Forest Service,* 46.

13. The two principal biographical studies of Pinchot are McGeary, *Gifford Pinchot,* and Harold T. Pinkett, *Gifford Pinchot: Private and Public Forester* (Urbana: University of Illinois Press, 1970). Pinchot's memoirs, *Breaking New Ground,* are far better than the average for that genre. See also the biographical account in John G. Waite and others, *Grey Towers: Preliminary Historic Structure Report,* prepared under contract for the Forest Service (Albany, N.Y.: Preservation/Design Group, 1978), 1–9, especially for the origins of his social consciousness. Pinchot was hired at Biltmore on the recommendation of his father's friend Frederick Law Olmsted, who conceived the idea of America's first managed forest as part of the estate's landscaping plan. Pinchot was fired in 1910 for defying presidential orders to cease making public castigations of the character of the secretary of the Interior in issues arising from the disposition of coal lands in Alaska.

14. (Samuel Trask Dana), "Editorial: Gifford Pinchot—Eighty Years Young," *JF* 43 (Aug. 1945): 547–48.

15. The summary of Cary's career, and the ensuing one of Schenck's, follow the biographical sketches in David A. Clary, "'Different Men from What We Were': Postwar Letters of Carl A. Schenck and Austin F. Cary," *JFH* 22 (Oct. 1978): 228–34. The only biography of Cary is Roy Ring White's "Austin Cary and Forestry in the South" (Ph.D. diss., University of Florida, 1961); see also White's "Austin Cary: The Father of Southern Forestry," *FH* 5 (Spring 1961): 2–5.

16. See note 15. There are no biographies of Schenck, and many of the available brief pieces are contradictory on some details. Among Schenck's many works see, in particular, *The Birth of Forestry in America: Biltmore Forest School, 1898–1913* (Santa Cruz, Calif.: FHS and the Appalachian Consortium, 1974), originally published as *The Biltmore Story: Recollections of the Beginning of Forestry in the United States* (St. Paul: Forest History Foundation and Minnesota Historical

Society, 1955). The school went wandering in search of a home after a dispute with Vanderbilt in 1909; it finally disbanded in 1913. Schenck did not vanish totally from the American scene, incidentally. He returned to the United States in 1924, and for the next few years he returned often to lecture and to teach at the University of Montana and occasionally at the Pennsylvania State Forest School. Largely retired during the Hitler years, he served American occupation forces as the chief forester of Greater Hesse after the war. Schenck's last official visit to the United States in 1951, old disputes forgotten, became a triumphal tour, with national news coverage (see "Forest Giant," *Newsweek,* 19 July 1951). On Schenck's work in Germany after World War II see Edward Stuart, Jr., "German Forestry during the American Occupation: Dr. Schenck's Pivotal Role," *JFH* 29 (Oct. 1985): 169–74.

17. J. E. Defebaugh, "The Changed Attitude of Lumbermen toward Forestry," *Proceedings of the American Forest Congress, 1905* (Washington, D.C.: H. M. Suter, 1905), 112.

18. "While this development of the agricultural interests of the Middle West has been in progress, the exploitation of the pine woods of the north has furnished another contribution to the commerce of the province. . . . As the white pine vanishes before the organized forces of exploitation, the remaining hard woods serve to establish factories in the former mill towns. The more fertile denuded lands of the north are now receiving settlers who repeat the old pioneer life among the stumps" (Frederick Jackson Turner, "The Middle West," in his *Frontier in American History,* 151–52).

19. On the history of the lumberman's frontier see Thomas Cox, "The Lumberman's Frontier," and the succession of regional articles on the lumber industry by different authors, together with their bibliographies, in Davis, ed., *Encyclopedia of American Forest and Conservation History.* The statistics are from *Statistical History of the United States,* 539–40.

20. Progressivism has inspired a vast literature and is periodically reinterpreted by historians. Works that are especially pertinent to its manifestation in conservation include Samuel P. Hays, *Conservation and the Gospel of Efficiency: The Progressive Conservation Movement, 1890–1920* (Cambridge: Harvard University Press, 1959), and James L. Penick, Jr., *Progressive Politics and Conservation: The Ballinger-Pinchot Affair* (Chicago: University of Chicago Press, 1968); see also McGeary, *Gifford Pinchot;* Elmo R. Richardson, "The Politics of the Conservation Issue in the Far West, 1896–1913" (Ph.D. diss., University of California at Los Angeles, 1958) and *The Politics of Conservation: Crusades and Controversies, 1897–1913* (Berkeley: University of California Press, 1962).

21. Hays's *Conservation and the Gospel of Efficiency* is the outstanding development of the technocratic thesis in Progressivism.

22. The Fabian conservation philosophies of these three men are discussed in a general way in Clary, "'Different Men.'" Schenck offered his views in *Birth of Forestry,* and Cary's are addressed in White, "Austin Cary and Forestry in the South." Greeley's own memoirs are *Forests and Men* (Garden City, N.Y.: Doubleday, 1951), as much an interpretation of conservation history as autobiography. All three were prolific writers who filled the professional journals and library bookshelves with their views, unpopular as those sometimes were among the more ideological Progressives. The entire story of Forest Service–industry relations, revolving especially around the regulation issue, is a very involved one, which is covered well by Steen, *U.S. Forest Service,* passim.

23. J. E. Rhodes, "Forestry and the Lumber Business," *FQ* 9 (June 1911): 195, 200, 202. *Stumpage* is an unavoidable part of forestry jargon. It means standing timber (timber still "on the stump"); it also applies to the value of such timber.

24. Wallace W. Everett, "The Practical in Forestry," *F* 5 (Dec. 1899): 276–77.

25. N. W. McLeod, "The Lumberman's Interest in Forestry," *Proceedings of the American Forest Congress, 1905* (Washington, D.C.: H. M. Suter, 1905), 100–101.

26. Frederick Weyerhaeuser, "How the Timber Supply May Be Perpetuated," *American Lumberman,* 14 July 1909, 36.

27. Schenck, "Training of Professional Foresters," 106.

28. Fernow's and Schenck's statements appear in Schenck, "Training of Professional Foresters," 103–6; "The True Forestry," *Forest and Stream,* reprinted in *F* 6 (Feb. 1900): 42.

29. Gifford Pinchot, *Practical Assistance to Farmers, Lumbermen, and Others in Handling Forest Lands,* USDA, Division of Forestry circ. no. 21 (Washington, D.C.: GPO, 1898), and "Notes on Some Forest Problems," *F* 5 (May 1899): 112 (for first quotation); Ralph S. Hosmer and Eugene S. Bruce, *A Forest Working Plan for Township 40 . . . New York State Forest Preserve,* USDA, Division of Forestry Bulletin no. 30 (Washington, D.C.: GPO, 1901), 12 (for second quotation).

30. Ford-Robertson, *Terminology,* 267.

31. Gifford Pinchot, *Practical Forestry,* pt. 2: *A Primer of Forestry,* USDA, Farmer's Bulletin no. 358 (Washington, D.C.: GPO, 1901), 12.

32. Steen, in *U.S. Forest Service,* 53–55, offers a good summary of the Circular 21 program.

33. Pinchot, *Practical Forestry,* 3; E. T. Allen, *Practical Forestry in the Pacific Northwest* (Portland, Oreg.: Western Forestry and Conservation Association, 1911), 11.

34. R. L. McCormick, "The Exhaustion of the Lumber Supply," *Forestry and the Lumber Supply,* USDA, Bureau of Forestry circ. no. 25 (Washington, D.C.: GPO, 1903), 8; R. C. Bryant, "The War and the Lumber Industry," *JF* 17 (Feb. 1919): 131.

35. Theodore Roosevelt, "Forestry and Foresters," address delivered before the Society of American Foresters, 26 Mar. 1903; reprinted in *Forestry and the Lumber Supply,* 4–5.

36. James Wilson to Gifford Pinchot, 1 Feb. 1905. This letter has been reprinted and distributed periodically by the Forest Service, and most employees are conversant with its tenor. The full text may be found in several places, including USDA, FS, *The Principal Laws Relating to Forest Service Activities,* USDA, Agricultural Handbook no. 453 (Washington, D.C.: GPO, 1978), 138–39. Pinchot quotes it in *Breaking New Ground,* 261–62, and credits its basic philosophy to W J McGee (p. 159; McGee, incidentally, never used periods after his initials).

37. Greeley, *Forests and Men,* 66. Political philosopher Jeremy Bentham (1748–1832) said that "The greatest happiness of the greatest number is the foundation of morals and legislation."

38. Samuel Trask Dana, *Forestry and Community Development,* USDA Bulletin no. 638 (Washington, D.C.: GPO, 1918), 3–5, 9–19; Elwood R. Maunder and Amelia Fry with Samuel Trask Dana, "Samuel Trask Dana: The Early Years," excerpts from oral history interviews, *FH* 10 (July 1966): 5. William B. Greeley offered similar findings in *Some Public and Economic Aspects of the Lumber Industry,*

pt. 1: *Studies of the Lumber Industry,* USDA Report no. 114 (Washington, D.C.: GPO, 1917), 90. Dana's report was published almost a decade after he prepared it.

39. Dana, *Forestry and Community Development,* 20–21.

40. Benton MacKaye, *Employment and Natural Resources* (Washington, D.C.: GPO, 1919), 21–23.

41. Pinchot, *Breaking New Ground,* 507 (for first quotation); USDA, Bureau of Forestry, *The Use of the National Forest Reserves* (Washington, D.C.: GPO, 1905), 32 (for second quotation; cited hereafter as *Use Book 1905*); USDA, FS, *The Use of the National Forests* (Washington, D.C.: GPO, 1907), 12 (for last quotation; cited hereafter as *Use Book 1907*).

42. ''Timber Sale Policy—District 5'' (n.d., c. 1908), pp. 9–13, in file Timber Sales, 1907–1911, R5-Buck. On the lumber industry on the West Coast see Thomas R. Cox, *Mills and Markets: A History of the Pacific Coast Lumber Industry to 1900* (Seattle: University of Washington Press, 1974) and ''Lumber Industry: Pacific Coast,'' in Davis, ed., *Encyclopedia of Forest and Conservation History,* 1:378–81; John H. Cox, ''Trade Associations in the Lumber Industry of the Pacific Northwest, 1899–1914,'' *Pacific Northwest Quarterly* 41 (Oct. 1950): 285–311; Edmond S. Meany, Jr., ''The History of the Lumber Industry in the Pacific Northwest to 1917'' (Ph.D. diss., Harvard University, 1936); Howard Brett Melendy, ''One Hundred Years of the Redwood Lumber Industry'' (Ph.D. diss., Stanford University, 1953); and Ralph W. Hidy, Frank Ernest Hill, and Allan Nevins, *Timber and Men: The Weyerhaeuser Story* (New York: Macmillan, 1963).

43. Austin Cary, ''Timber Sale Business in California,'' Mar. 1914, with accompanying memorandum dated Sept. 1913, S-Studies, Reports by Austin Cary, 1912–1914, 54-A-111/59858, NARS-S.

44. U.S., Department of Commerce, Bureau of Corporations, Report on the Lumber Industry, xxii. The bureau became the Federal Trade Commission in 1915.

45. The situation of the national forests was considerably complicated by persistent attempts to rescind the reserves after they were established, attempts that were often born out of a misunderstanding that the lands and resources on the reserves would thereafter be unavailable for use. There was also a considerable element of East/West sectionalism in every public dispute involving the forests, which reflected longstanding western complaints that eastern-dominated government restricted westerners' opportunities and gave nothing in return. The history of the public lands, and of their disposal, is traced well in Roy M. Robbins, *Our Landed Heritage: The Public Domain, 1776–1936* (Lincoln: University of Nebraska Press, 1942), and in Everett Dick, *The Lure of the Land: A Social History of the Public Lands from the Articles of Confederation to the New Deal* (Lincoln: University of Nebraska Press, 1970). Other complications were introduced by Congress. The Organic Act of 1897 included a provision for the exchange of ''lieu lands,'' something that Henry S. Graves called ''probably one of the worst sources of looting the public lands in their whole history'' (Graves to the DF, San Francisco, 13 July 1914, USFS, Dist. 4, S-Supervision, Non-statistical Reports, 1913–1916, 60-A-192/601106, NARS-D). The measure allowed holders of unperfected claims or patented lands to exchange their tracts within forest-reserve boundaries for public lands in other locations. It was a license to steal, and although Congress repealed the provision five years before the Forest Service was created, it left an odor that tainted legitimate land exchanges. It also

made forest-reserve boundaries appear easily adjustable (see Dana, *Forest and Range Policy,* 113). The Forest Homestead Act of 11 June 1906 (24 Stat. 233) scrambled affairs further. It permitted the return or exchange of land to the public domain when it proved to be either unsuitable for timber harvest or best devoted to agriculture. The criteria were questionable, and the Forest Service was bemused, so much so that after interminable debate, forest supervisors found themselves unable to respond to applications for land exchanges (see "Report of Supervisors' Meeting at Missoula, Montana," *FQ* 8 [Sept. 1910]: 321–22).

46. U.S., Department of the Interior, *Forest Reserve Manual* (Washington, D.C.: GPO, 1902), 3; *Use Book 1905,* 4 (emphasis added); *Use Book 1907,* 7.

47. 30 Stat. 34; Pinchot, *Breaking New Ground,* 276.

48. *Use Book 1905,* 7; *Use Book 1907,* 17.

49. 34 Stat. 669, 684; 35 Stat. 260. The monies were to be spent by the counties, as prescribed by the state legislatures, on public schools and roads in the respective counties.

50. Edward A. Braniff, "The Reserve Policy in Operation," *FQ* 2 (May 1904): 141; McGeary, *Gifford Pinchot,* 93.

51. Lincoln National Forest, *The Alamo Adviser,* Dec. 1910, 2.

52. George H. Cecil to the Forester, 6 May 1911, USFS, Dist. 1, S & ST-District, Policy 1911–1913, R1, 63-A-209/82498, NARS-S.

53. "Report of Supervisors' Meeting at Missoula," 303.

CHAPTER TWO. FORGING THE TIMBER-MANAGEMENT PROGRAM

1. U.S., Department of the Interior, *Rules and Regulations Governing Forest Reserves* (Washington, D.C.: GPO, 1897; hereafter cited as Interior, *Rules and Regulations 1897*).

2. Ibid.; U.S., Department of the Interior, *Rules and Regulations Governing Forest Reserves* (Washington, D.C.: GPO, 1900), 10–11.

3. J. W. Dobson, forest supervisor, to commissioner, General Land Office, 28 July 1899, and C. M. Peckinpah and T. E. Peckinpah (purchasers) to commissioner, General Land Office, 12 May 1899, Timber Sales, 1907–1911, R5-Buck.

4. Interior, *Forest Reserve Manual,* 36n.

5. *Use Book 1905,* 43.

6. *Statistical History of the United States,* 534; USDA, FS, Dist. 5, "Timber Sale Policy—District 5" (n.d.), 20, Timber Sales, 1907–1911, R5-Buck.

7. E. E. Carter to DF [May 1908], and Overton Price to E. A. Sherman, 2 June 1908, S-Sales, Stumpage Prices, 1908–1914, 1916, R1, 63-A-209/82493, NARS-S.

8. Chief of Silviculture, Dist. 5, to forest supervisors, 3 Mar. 1909, Appraisal Methods, 1909–1913, R5-Buck.

9. E. A. Sherman, "Standard and Minimum Stumpage Prices Recommended for the National Forests in District No. 1," attached to Sherman to the Forester, 20 Oct. 1908, S & ST-District, Policy, 1908-1910, R1, 63-A-209/82498, NARS-S; Dist. 1 to the Forester, 19 Feb. 1909, S-Sales, Stumpage Prices, 1908–1914, 1916, R1, 63-A-209/82493, NARS-S.

10. William T. Cox to DF, 11 June 1909, and F. A. Silcox to the Forester, 26 Feb. 1910, S-Sales, Stumpage Prices, 1908–1914, 1916, R1, 63-A-209/82493, NARS-S.

11. E. E. Carter to DF, 22 Mar. 1910, S-Sales, Stumpage Prices, 1908–1914, 1916, R1, 63-A-209/82493, NARS-S.

12. Interior, *Rules and Regulations 1897,* 11, 9.

13. USDA, FS, *The Forest Service: What It Is and How It Deals with Forest Problems,* USFS circ. no. 26, 2d ed. (Washington, D.C.: GPO, 1906), 22; William T. Cox to DF, 22 Feb. 1910, S & ST-District, Policy, 1908–1910, R1, 63-A-209/82498, NARS-S. ''Definite'' quantities were defined as ''average conditions for the whole sale.''

14. *Statistical History of the United States,* 534, 539.

15. Cox to DF, 11 Feb. 1910; USDA, FS, ''Silviculture Management'' (1910), 4; Swift Berry to forest supervisors, Dist. 5; and USDA, FS, Dist. 5, ''Notes and Policy: Silviculture,'' Feb. 1910, p. 2—all in Timber Sales, 1907–1911, R5-Buck. The last-named document said that the decision reflected the ''wish'' of the Agriculture Committee of the House of Representatives.

16. F. A. Silcox to forest supervisors, 10 Mar. 1910, S & ST-District, Policy, 1908–1910, R1, 63-A-209/82498, NARS-S.

17. Graves to DF, 21 Mar. 1910, S & ST-District, Policy, 1908–1910, R1, 63-A-209/82498, NARS-S.

18. W. B. Greeley to the Forester, 31 Mar. 1910, S & ST-District, Policy, 1908–1910, R1, 63-A-209/82498, NARS-S.

19. ''Supervisors' Meeting at San Francisco,'' *FQ* 9 (Mar. 1911): 69.

20. ''Timber Sale Policy—District 5'' (n.d.), and ''Estimate and Timber Sale Policy Data—District 5'' (n.d., received in Dist. 5 on 6 July 1908), Timber Sales, 1907–1911, R5-Buck.

21. A. B. Recknagel, ''The Progress of Reconnaissance,'' *FQ* 8 (Dec. 1910): 417.

22. *Statistical History of the United States,* 538; W. B. Greeley to the Forester, 29 Aug. 1911, S & ST-District, Policy, 1911–1913, R1, 63-A-209/82498, NARS-S.

23. Cary to Greeley, 16 Sept. 1911, S & ST-District, Policy, 1911–1913, R1, 63-A-09/82498, NARS-S.

24. Burt P. Kirkland, ''The Need of a Vigorous Policy of Encouraging Cutting on the National Forests of the Pacific Coast,'' *FQ* 9 (Sept. 1911): 379–89.

25. Graves to DF, 26 May 1911, S & ST-District, Policy, 1911–1913, R1, 63-A-209/82498, NARS-S; Graves to DF, 13 July 1914, S-Supervision, Nonstatistical Reports, 1913–1916, R4, 60-A-192/601106, NARS-D.

26. Graves to DF, 13 July 1914, cited in note 25.

27. E. A. Sherman, ''The Sawmill of the Future,'' *FQ* 4 (Dec. 1908): 369; ''News and Notes,'' *JF* 8 (Sept. 1910): 400–401 (billion BF sale); USDA, FS, *Forest Service Manual* (FSM-1-1-1911), ''Timber Sales,'' amendment 163, 21 Mar. 1913. The large timber sale was an inducement to the Humboldt and Eastern Railway to build a line from Eureka to the Sacramento Valley in California.

28. O. M. Butler, ''District Market Plan—District 4'' (typescript, 71 pp., Jan. 1912, cited hereafter as Butler, ''D4 Market Plan—1912''); and E. A. Sherman to forest supervisor, 1 May 1913 (for last quotation), both in S-Supervision, General, 1912–1915, R4, 60-A-192/601106, NARS-D.

29. Graves to DF, 26 May 1911, cited in note 25.

30. Graves to DF, Dist. 5, 8 Jan. 1912, Appraisal Methods, 1909–1913, R5-Buck.

31. C. S. S. [Smith?], Office of Products, Dist. 5, ''Memorandum for D and S,'' 12 Aug. 1913, Timber Sales, 1912, R5-Buck. District 5 was thinking about selling no more than was necessary to keep established plants in operation.

32. George H. Cecil to the Forester, 1 June 1912, S-Supervision, General, 1912–1915, R4, 60-A-192/601106, NARS-D.

33. Greeley to DF, Dist. 6, 4 Sept. 1912, S-Supervision, General, 1912–1915, R4, 60-A-192/601106, NARS-D.

34. Graves to DF, 6 Feb. 1913, S-Sales, Stumpage Appraisals, 1907–1915, R1, 63-A-209/82491, NARS-S.

35. Smith to the DF, 10 Oct. 1911, Silcox to the Forester, 19 Oct. 1911, and Graves to Smith, 6 Nov. 1911—all in S & ST-District, Policy, 1911–1913, R1, 63-A-209/82498, NARS-S.

36. Graves to DF, 6 Nov. 1911, S & ST-District, Policy, 1911–1913, R1, 63-A-209/82498, NARS-S; Swift Berry, "Timber Sales [D5]," Timber Sales, FHS, Santa Cruz, Calif.; and USDA, FS, *Sale Prospectus: Crater and Paulina National Forests, Oregon* (Portland, Oreg.: USDA, FS, 1915).

37. Interior, *Rules and Regulations 1897,* 8; *Use Book 1907,* 18; Pinchot, *Breaking New Ground,* 140.

38. USDA, FS, *Forest Service,* 10.

39. *Use Book 1907,* 9; USDA, FS, *Forest Service,* 2–21; George W. Tucker, "Forest Service in the Southwest," 1343, 1680 History File, Southwest Regional Office, Albuquerque, N. Mex.

40. "Fishermen for the Forests," *F* 5 (June 1899): 125.

41. Interior, *Forest Reserve Manual 1902,* 56, 58; *Use Book 1907,* 16 (for quotation).

42. *Statistical History of the United States,* 534, 539.

43. USDA, FS, *The Timber Supply of the United States,* circ. no. 97, comp. Royal S. Kellogg (Washington, D.C.: GPO, 1907), 4; USDA, FS, *American Forest and Forest Products* (Washington, D.C.: GPO, 1927), 297, 299, 300; "News and Notes," *FQ* 6 (Sept. 1908): 300; *Statistical History of the United States,* 539.

44. "The Forests of the United States," *F* 6 (Mar. 1900): 55.

45. "Supervisors' Meeting, District 5," *FQ* 9 (Mar. 1911): 69.

46. Recknagel, "Progress of Reconnaissance," 415; R. E. Marsh, "Timber Cruising," *FH* 13 (Oct. 1969): 22.

47. [W. B. Greeley], "Report of Supervisors' Meeting at Missoula, Montana," *FQ* 8 (Sept. 1910): 308. For other methods see Graves, *Woodsman's Handbook,* 120; Recknagel, "Progress of Reconnaissance," 417; and Will Mace to Gifford Pinchot, 9 Feb. 1940, 1680 History Files, Uncle Jim Owens, Kaibab NF.

48. Aldo Leopold to DF, 1 Nov. 1909, S-Sales, Apache, Blue River Wagon Road Project, Historical Files, Apache NF.

49. USFS District 1, *Manual for Timber Reconnaissance* (Missoula, Mont.: USDA, FS, Dist. 1, 1914), 8–14, 31–33.

50. Recknagel, "Progress of Reconnaissance," 417 (for "long-road" quotation); Graves, *Woodsman's Handbook,* 130–31.

51. USDA, Division of Forestry, *Practical Assistance;* Graves, *Practical Forestry;* Henry S. Graves, "Conservation and Restoration," *F* 5 (July 1899): 161; E. M. Griffith, "The Lumberman's Point of View: Conservation," *F* 5 (June 1899): 134–35; Henry S. Graves, "Practice of Forestry by Private Owners," in USDA, *Yearbook 1899* (Washington, D.C.: GPO, 1900), 419–20; Overton W. Price, "Studying the Adirondack Forest," *F* 6 (Jan. 1900): 19; E. M. Griffith, "Forest Development," *F* 6 (Jan. 1900): 18; T. R. Sherrard, "Forest Work on the Pacific Slope," *F* 6 (Jan. 1900): 20; "Investigations in North Carolina," *F* 6 (Jan. 1900): 36; "A Study of Hemlock," *F* 6 (Apr. 1900): 89; and "Studying American Forests," *F* 6 (Apr. 1900): 88. Circular-21 crews surveyed species, growth, and

reproduction on each of the lands visited. With those data in hand, they offered recommendations on tree-size limits for cutting, rules for marking, and cutting levels and practices for sustained production and profits. They sometimes even suggested reprimands for violations of the working plan.

52. D. T. Mason, "Reconnaissance and Working Plans," in [W. B. Greeley], "Report of Supervisors' Meeting," 307–9.

53. Greeley to DF, Dist. 6, 4 Sept. 1912, and Cecil to the Forester, 6 May 1911, S & ST-District, Policy, 1911–1913, R1, 63-A-209/82498, NARS-S.

54. Butler, "D4 Market Plan—1912."

55. Cecil to the Forester, 6 May 1911, cited in note 53.

56. Pinchot, *Primer of Forestry,* 7–8.

57. Early methods—according to B. A. Chandler, "Notes on a Method of Studying Current Growth Percent," *FQ* 24 (Sept. 1916): 453–60—were originated by H. H. Chapman in 1909 and expanded by J. G. Stetson in "Suggestions on Predicting Growth," *FQ* 9 (Sept. 1910): 326. Hanzlik's "site quality factor" exercise is described in great detail in E. J. Hanzlik, "The Growth and Yield of Douglas Fir on Various Sites in Western Washington and Oregon," 14 Mar. 1912, typed, 35 pp., S-Studies, Growth and Yield of Douglas Fir . . . E. J. Hanzlik, 1912, R6, 54-A-111/59859, NARS-S. All formulas, theories, and calculations discussed in this chapter are presented in complete form, with further discussion, in David A. Clary, *More Wood for the Nation: A History of Timber Management on the National Forest System of the United States,* report DAC-10 (Bloomington, Ind.: David A. Clary & Associates, 1983), available in the files of the History Section, USDA, FS, Washington, D.C.

58. Graves, *Woodsman's Handbook,* 10, 14–17; A. L. Daniels, "The Measurement of Saw Logs and Round Timber," *FQ* 3 (Nov. 1905): 339. Variations were also true of log rules.

59. Graves, *Woodsman's Handbook,* 100–106.

60. E. T. Allen, "Red Fir" (1903), mimeographed, 94 pp., S-Studies, Red Fir, E. T. Allen, 1903, R6, 54-A-111/59862, NARS-S. From his data, Allen concluded that volume, as calculated by the Scribner Decimal C rule, was "fair," if "used at the center of the log as was the intention of the inventor," but that "the universal practice of scaling at the small end undervalues large logs to an extent seldom realized by the lumbermen." As a result of his work and that of others, the Forest Service adopted the Scribner Decimal C rule—which calculated the volume of a 16-foot log—for measuring the content of logs.

61. Judson F. Clark, "The Measurement of Saw Logs," *FQ* 4 (June 1906): 80. Clark was the inventor of the International Log Rule.

62. Louis Margolin, "Mill Scale Studies," *FQ* 4 (Mar. 1906): 5–7.

63. J. P. Hughes to the Forester, through S. C. Bartrum, 9 Mar. 1908, F. A. Silcox to DF, Portland, Oreg., 20 Sept. 1911, and George H. Cecil to DF, Missoula, Mont., 7 Oct. 1911—all in S & ST-District, Policy, 1911–1913, R1, 63-A-209/82498, NARS-S.

64. A typical example is H. B. Oakleaf, "Report on Mill Scale Studies at Ketchikan and Petersburg, Alaska," typed, 22 pp. with 2 supplements, S-Studies, Logging and Milling, 1904–1913, R6, 54-A-111/59860, NARS-S. Another factor that affected timber-management planning and the effort to achieve sustained yield was the passage of the Weeks Act, which took effect in 1911. It authorized the government to purchase lands for timber and watershed management (36 Stat. 961-963, as amended).

65. European cutting systems were discussed in a number of publications: see, e.g., Pinchot, *Primer of Forestry,* 15–18, and Fernow, *Economics of Forestry,* 173–75. The coppice system is not discussed in this text because the Forest Service did not use it.

66. Graves, ''Practice of Forestry,'' 420–21.

67. Hosmer and Bruce, *Forest Working Plan,* 10, 33, 38; Graves, *Practical Forestry,* 14.

68. Pinchot, *Breaking New Ground,* 186; ''An Expert Opinion on the Cornell College Forest Experiment,'' *FQ* 3 (Feb. 1905): 32–38; see also ''Three State Bills,'' *FQ* 1 (July 1903): 156; ''Report of the New York State College of Forestry,'' *FQ* 1 (Apr. 1903): 103; ''Current Literature and Reviews,'' *FQ* 1 (Apr. 1903): 104; ''The Collapse of the New York State College of Forestry,'' *FQ* 2 (Nov. 1903): 44; B. E. Fernow, ''The Results of Systematic Forest Management,'' *FQ* 6 (Sept. 1908): 229–36; and E. A. Sterling, ''Artificial Reproduction of Forests,'' *FQ* 6 (Sept. 1908): 211.

69. ''Expert Opinion,'' 35; Pinchot, *Breaking New Ground,* 151.

70. Allen, ''Red Fir,'' 1. He was correct. It has been the preeminent lumber species since World War II (*Statistical History of the United States,* 543).

71. Allen, ''Red Fir,'' 2, 21–27, 76, 86–88. Anson E. Cohoon confirmed Allen's findings in ''Silvical Report on the Cascade (North) Forest Reserve, Oregon'' (1906), typed, 47 pp., S-Studies, Willamette (Cascade North) Forest Reserve Annual Silvical Report, A. E. Cohoon, 1906, R6, 54-A-111/59826, NARS-S.

72. Chester B. Cox to Clyde R. Seitz, forest supervisor, Cascade NF, 25 Mar. 1909, S-Studies, Willamette (Cascade) Forest Annual Silvical Report, C. B. Cox, 1909, R6, 54-A-111/59862, NARS-S; ''A Supervisors' Meeting,'' *FQ* 8 (June 1910): 215.

73. George H. Cecil to the Forester, 6 May 1911, S & ST-District, Policy, 1911–1913, R1, 63-A-209/82498, NARS-S.

74. Greeley to the Forester, 29 Aug. 1911, and Cary to Greeley, 6 Sept. 1911, S & ST-District, Policy, 1911–1913, R1, 63-A-209/82498, NARS-S; Kirkland, ''Need of a Vigorous Policy,'' 376–77; George Y. Baker, ''Silvical Study of Stands in the Willamette Watershed,'' Jan. 1914, typed, 13 pp., S-Studies, Stands in the Willamette Watershed, R6, 54-A-111/59862, NARS-S.

75. David T. Mason, with Elwood R. Maunder, ''Memoirs of a Forester,'' *FH* 10 (Jan. 1967): 31.

76. ''Supervisors' Meeting [Missoula, Mont.],'' 305–6.

77. William H. Kobbe, ''Annual Silvical Report for the Apache National Forest,'' 3 Apr. 1909; William H. Kobbe, ''Annual Silvical Report for the Apache National Forest,'' 3 Apr. 1911; and Aldo Leopold to district forester, dist. 3, 20 Mar. 1910, all in S-Plans, Annual Silvicultural Report, 1909, Apache National Forest Historical Files, Ariz.; Cecil to the Forester, 6 May 1911, and Cox to Seitz, 25 Mar. 1909, cited in notes 73 and 72.

78. USFS, ''Silviculture Management,'' Feb. 1910, cited in note 15.

79. Frank J. Harmon, ''Remembering Franklin B. Hough,'' USFS reprint from *American Forests* (Jan. 1977): 53; Fernow, *Economics of Forestry;* Richard A. Overfield, ''Trees for the Great Plains,'' *JFH* 23 (Jan. 1979): 18–31.

80. E. A. Sterling, ''Preliminary Plan for Forest Experiments in Cooperation with the Madera Sugar Pine Company,'' Aug. 1904, typed, 8 pp., (T) Reforestation (Nursery Work), TMO, Sequoia NF, Calif.

81. Pinchot, *Breaking New Ground,* 309; see also Raphael Zon, "Plan for Creating Forest Experiment Stations," 6 May 1908, Research Compilation File, RG95, USFS, NA.

82. G. A. Pearson, "The Oldest Forest Experiment Station," 13 Apr. 1936, in Tucker, "Forest Service in the Southwest," 1335.

83. *Plateau* 28 (Apr. 1956): 86–87; Pearson, "Oldest Forest Experiment Station," 1336–37; G. A. Pearson, *Ponderosa Pine in the Southwest* (Washington, D.C.: GPO, n.d.).

84. David S. Olson, "The Savenac Nursery," *FH* 11 (July 1967): 15; USFS, "Silviculture Management," 7; William T. Cox to DF, Dist. 1, 6 Feb. 1909, S & ST-District, Policy, 1908–1910, R1, 63-A-209/82498, NARS-S.

85. *Plateau* 28 (Apr. 1956): 87.

86. Among many general studies of the various factors involved here are three superb pieces of scholarship: Stephen J. Pyne, *Fire in America: A Cultural History of Wildland and Rural Fire* (Princeton, N.J.: Princeton University Press, 1982); Susan R. Schrepfer, *The Fight to Save the Redwoods: A History of Environmental Reform, 1917–1978* (Madison: University of Wisconsin Press, 1983); and Susan L. Flader, *Thinking Like a Mountain: Aldo Leopold and the Evolution of an Ecological Attitude toward Deer, Wolves, and Forests* (Columbia: University of Missouri Press, 1974). Each is far broader in coverage and outlook than its title suggests, and each is thoroughly grounded in an understanding of the evolution of scientific wisdom regarding forest environments. On myth, science, and the Forest Service see Ashley Leo Schiff, *Fire and Water: Scientific Heresy in the Forest Service* (Cambridge: Harvard University Press, 1962), the contents of which bear out the subtitle.

87. Interior, *Forest Reserve Manual, 1902,* 41; S. J. Flintham, "Forest Extension in the Sierra Forest Reserve" (1904), typed, 203 pp., History File, PIO, Sierra NF, Calif.

88. Cox to DF, 6 Feb. 1909, cited in note 84; "News and Notes," *FQ* 6 (Sept. 1908): 319–22.

89. "Supervisors' Meeting—District 5," 71; W. B. Greeley to DF, Ogden, Utah, 2 May 1912, and R. Y. Stuart to DF, San Francisco, Calif., 14 Mar. 1914, Eldorado, Corrrespondence, 1907–1918, R5, 54-A-760/24725, NARS-San Bruno (NARS-SB).

90. *Statistical History of the United States,* 535.

91. John H. Hatton to forest supervisor, Dist. 5, 26 Feb. 1912, Timber Sales, 1912, and DuBois to the Forester, 28 Nov. 1911, Appraisal Methods, 1909–1913—both in R5-Buck. A good history of Forest Service timber appraisals is in Alfred A. Wiener's *The Forest Service Timber Appraisal System: A Historical Perspective, 1891–1981,* USDA, FS publication FS-381 (Washington, D.C.: USDA, FS, 1982).

92. E. A. Sherman to forest supervisors, 1 May 1913, S-Supervision, General, 1912–1915, R4, 60-A-192/601106, NARS-D.

93. James W. Girard, "Forest Service Stumpage Appraisals," *JF* 15 (Oct. 1917): 708; P. V. L. [?], "Memo for Fromme," " 'History of Forest,' " 1912, Timber Sales—Historical Data, R. L. Fromme, 1911–1962, PIO, Olympic NF, Olympia, Wash.

94. Swift Berry, "Timber Sales," Jan. 1912, Timber Sales, R5-Buck; Cary, "Timber Sale Business in California," cited in note 43 of chap. 1; Butler, "D4 Market Plan—1912," 58.

95. Butler, "D4 Market Plan—1912," 47.

96. Chapman, *Forest Valuation,* v; Fernow, *Economics of Forestry,* 215, 251.

97. Allen, ''Red Fir.''

98. R. W. Ayers, *History of Timber Management in the California National Forests, 1850–1937* (Washington, D.C.: USDA, FS, 1958), 17; Chapman, *Forest Valuation*, v. I did not discover any copy of Schenck's pamphlet while I was doing research for this study.

99. District 5, ''Timber Sale Policy,'' cited in note 6.

100. Cox to DFs, 11 Feb. 1910, and USDA, FS, ''Silvicultural Management—Notes on Policy,'' Feb. 1910, S & ST-District, Policy 1908–1910, R1, 63-A-209/82498, NARS-D. In District 5, this was sent to the national forests as ''Notes on Policy: Silviculture,'' Feb. 1910 (see Timber Sales, 1907–1911, R5-Buck).

101. DuBois to forest supervisors, 3 Aug. 1911, Appraisal Methods, 1909–1913, R5-Buck.

102. DuBois to the Forester, 18 Nov. 1911, Appraisal Methods, 1909–1913, R5-Buck.

103. Graves to DF, Dist. 5, 8 Jan. 1912, Appraisal Methods, 1909–1913, R5-Buck.

104. Woodbury (dictated by DuBois) to the Forester, 10 Feb. 1912; and Swift Berry, ''Memorandum for District Forester,'' 3 Feb. 1912, Appraisal Methods, 1909–1913, R5-Buck.

105. Graves to DF, Dist. 5, 28 Feb. 1912, Appraisal Methods, 1909–1913, R5-Buck. Graves had several other objections to Berry's formula, or what he called ''the average total investment'' plan, as against the Forester's formula, or ''operating profit'' plan.

106. Butler, ''D4 Market Plan—1912,'' 62–63, 67–71.

107. Graves to DF, Dist. 4, 1 Apr. 1912, S-Sales, Stumpage Appraisals, 1907–1915, R1, 63-A-209/82491, NARS-S.

108. Dorr Skeels, ''Memorandum for District Forester,'' 11 Mar. 1912, S-Supervision, General, 1912–1915, R4, 60-A-192/601106, NARS-D.

109. Silcox to the Forester, 3 Apr. 1912, ibid.

110. Greeley to DF, Dist. 1, 1 May 1912, ibid.

111. William B. Hunter, ''The Valuation of Stumpage,'' 5 May 1912, S-Studies, Western Callam and Jefferson Counties, R6, 54-A-111/59862, NARS-S. Hunter determined ''X'' by a simple formula involving costs and profit rates.

112. Unfortunately, none of these contributions survives in the records. Numerous references, however, cite their existence: see Greeley to DF, Dist. 6, 4 Sept. 1912; Cecil to the Forester, 31 May 1912, and Cecil to the Forester, 1 June 1912—all in S-Supervision, General, 1912–1915, R4, 60-A-192/601106, NARS-D.

113. Greeley to DF, 4 Sept. 1912, ibid.

114. Greeley to the Forester, 19 Aug. 1911, S & ST-District, Policy, 1911–1913, R1, 63-A-209/82498, NARS-S; and for a similar opinion, DuBois to the Forester, 28 Nov. 1911, Appraisal Methods, 1909–1913, R5-Buck.

115. Graves to DuBois, 8 Jan. 1912, and Graves to DF, Dist. 5, 9 Feb. 1912, Timber Sales, 1912, R5-Buck. District 6 had also advocated the use of lumber grades in market prices. Greeley disapproved that suggestion ''because of my distrust of their stability'' (see Greeley to district forester, Dist. 6, 4 Sept. 1912, cited in note 112). Graves's decision to average mill-lumber prices over two years was undoubtedly influenced by a study being completed by Austin Cary during 1912, which, among other things, called for three-year averages to benefit the government (Austin Cary, ''Relating Stumpage to Lumber Prices,'' 1912, with addendum, S-Studies, Reports by Austin Cary, 1912–1914, R6, 54-

A-111/59858, NARS-S. The front cover of Cary's report carries the date 1912, but Cary did not sign it until 5 June 1913. An additional addendum was added in January 1914.

116. Graves to DF, 26 May 1911, Greeley to the Forester, 29 Aug. 1911, and DuBois to the Forester, 28 Nov. 1911, cited in notes 29, 22, and 91, respectively; untitled and undated newspaper clipping, USFS Clipping Files, Timber Sales, FHS; Swift Berry, "Timber Sales," 19 Jan. 1912, Timber Sales, 1907–1911, R5-Buck; Graves to DFs, 11 Feb. 1913, S-Sales, Stumpage Appraisals, 1907–1915, R1, 63-A-209/82491, NARS-S. Graves had difficulty in getting his readjustment decision accepted in the Districts, and it was a subject of interminable arguments (see Graves to DF, 8 Jan. 1912, cited in note 103; Cecil to the Forester, 1 June 1912, cited in note 112; and Greeley to DF, Dist. 6, 4 Sept. 1912, cited in note 115; Silcox to the Forester, 6 Mar. 1912, S-Sales, Stumpage Appraisals, 1907–1915, R1, 63-A-209/82491, NARS-S; and Cary, "Timber Sales in Douglas Fir," 1913, pp. 49–54, S-Studies, Reports by Austin Cary, 1912–1914, 54-A-111/59858, NARS-S).

117. Graves to DF, 1 Apr. 1912; Greeley to DF, 4 Sept. 1912, cited in notes 107 and 115.

118. Graves to DF, Dist. 4, 1 Apr. 1912, cited in note 107. W. H. Gibbons, in District 6, thought there were several problems with the Forester's approach to residual value: "As a general proposition, it seems it would depend on how badly we wished to sell the timber." Instead, Gibbons suggested, it might be better to account for the operator's investment by finding another means of guaranteeing future sales to the first purchaser ("Memorandum for Silviculture," 10 Apr. 1912, S-Supervision, General, 1912–1915, R4, 60-A-192/601106, NARS-D). George Cecil also was displeased with the Forester's "inflexible" decision on depreciation and investment charges (Cecil to the Forester, 1 June 1912, cited in note 32).

119. Greeley to DF, Dist. 6, 4 Sept. 1912, cited in note 115.

120. DuBois to the Forester, 28 Nov. 1911, cited in note 114.

121. Graves to DuBois, 8 Jan. 1912, cited in note 115. These comments were passed on to the District 5 forests in Berry, "Timber Sales," 19 Jan. 1912, Timber Sales, 1907–1911, R5-Buck. A few months later, Graves set the minimum rate for inferior species at 50 cents per thousand board feet, "approximately the cost of administration of National Forest sales" (Graves to DF, Dist. 5, 9 Feb. 1912, and Greeley to DF, Dist. 6, 4 Sept. 1912, cited in notes 115 and 112).

122. Butler, "D4 Market Plan—1912," 47–48; Greeley to DF, Dist. 6, 6 Sept. 1912; Sherman to forest supervisors, 2 May 1913; and Graves to DFs, Ogden and Portland, 23 Jan. 1913—all in S-Supervision, General, 1912–1915, R4, 60-A-192/601102, NARS-D.

123. Greeley to DF, Dist. 6, 4 Sept. 1912, cited in note 112.

124. After a meeting with both Districts and after a review of the material, Greeley ruled in Portland's favor, leaving Ogden greatly miffed (Greeley to DFs, Ogden and Portland, 29 Aug. 1912; Hoyt to the Forester, 16 Sept. 1912; and Graves to DFs, Ogden and Portland, 23 Jan. 1913—all in S-Supervision, General, 1912–1915, R4, 60-A-192/601102, NARS-D).

125. H. H. Chapman, "Suggestions by H. H. Chapman on Method of Determining Stumpage Appraisals of National Forest Timber," July 1912, S-Sales, Stumpage Appraisals, 1907–1915, R1, 63-A-209/82491, NARS-S.

126. Cary, "Timber Sales in Douglas Fir," 1913, pp. 36–48.

127. Swift Berry, *Stumpage Appraisal Handbook, District 5* (San Francisco: USDA, FS, 1913), 10–14.

128. Sherman to forest supervisor, 1 May 1912, S-Supervision, General, 1912–1915, R4, 60-A-192/601106, NARS-S. The term *conversion costs* also appeared in Hunter's appraisal suggestions.

129. Greeley to DF, 1 Mar. 1914, S-Sales, Stumpage Appraisals, 1907–1915, R1, 63-A-209/82491, NARS-S. Greeley received responses from the Districts and sent them an updated copy six months later (Greeley to DF, 1 Sept. 1914, same file).

130. USDA, FS, *Instructions for Appraising Stumpage on National Forests* (Washington, D.C.: GPO, 1914), 7; cited hereafter as *Appraisal Instructions 1914.*

131. Ibid., 38–43.

132. Ibid., 9–10, 16–17, 35, 41, 44.

CHAPTER THREE. SELLING TIMBER IN AN UNCERTAIN MARKET

1. Steen, *U.S. Forest Service,* 113.

2. Ibid., 137–40; see also Schiff, *Fire and Water.*

3. Steen, *U.S. Forest Service,* 113–22. On the origins and early history of recreational development see William C. Tweed, *Recreation Site Planning and Improvement in National Forests, 1891–1942,* FS Publication FS-354 (Washington, D.C.: USDA, FS, 1980).

4. The summary of World War I that follows relies in great part on David A. Clary, "World War I and American Forests and Forest Industry," in Davis, ed., *Encyclopedia;* among others, see also these articles by Clary: "The Biggest Regiment in the Army" and "The Woodsmen of the AEF: A Bibliographical Note"; James E. Fickle, "Defense Mobilization in the Southern Pine Industry: The Experience of World War I"; and Daniel R. Mortensen, "The Deterioration of Forest Grazing Land: A Wider Context for the Effects of World War I"—all in *JFH* 22 (Oct. 1978). For an account of wartime construction, with emphasis on materials, see David A. Clary, *A Life Which Is Gregarious in the Extreme: A History of Furniture in Barracks, Hospitals, and Guardhouses of the United States Army, 1880–1945,* report prepared under contract for the NPS, Report DAC-9 (Bloomington, Ind.: David A. Clary & Associates, 1983), chap. 7.

5. Quoted in Steen, *U.S. Forest Service,* 143.

6. Quoted in Clary, "Biggest Regiment," 182; see also George T. Morgan, ed., "A Forester at War: Excerpts from the Diaries of Colonel William B. Greeley, 1917–1919," *JFH* 4 (Winter 1961).

7. Steen, *U.S. Forest Service,* 176–78; "Forest Devastation: A National Danger and a Plan to Meet It," *JF* 17 (Dec. 1919): 911–45; Henry Graves, "A Policy of Forestry for the Nation," *JF* 17 (Dec. 1919): 901–10.

8. *Hammer* v. *Dagenhart,* 247 U.S. 251 (1918), and *Bailey* v. *Drexel Furniture Company,* 259 U.S. 20 (1922). Other examples could be cited.

9. See, e.g., Austin Cary to Carl A. Schenck, 9 Dec. 1924, in Clary, "Different Men," 234.

10. Press releases, July and Nov. 1921 and 23 Mar. 1923, USFS, Clipping File, Timber Sales, FHS; Steen, *U.S. Forest Service,* 148–49.

11. Fred E. Ames, "Memorandum in Regard to History, Policy and Procedure Relating to Sales of Pulp Timber in British Columbia," Oct. 1921, S-Sales,

British Columbia Pulp and Paper, R-6, 54-A-111/59862, NARS-S; Steen, *U.S. Forest Service,* 148–49.

12. Steen, *U.S. Forest Service,* 150–52.

13. Greeley to DF, 18 June 1920, S-Legislation (1920–1934) Closed, R-6, 54-A-111/59854, NARS-S; Steen, *U.S. Forest Service,* 179–81.

14. 43 Stat. 653; Steen, *U.S. Forest Service,* 179–93. A recent administrative history of Forest Service cooperative programs, prepared under contract for the Forest Service, is William G. Robbins's *American Forestry: A History of National, State, and Private Cooperation* (Lincoln: University of Nebraska Press, 1985).

15. Tucker, "Forest Service in the Southwest" (cited in note 39 of chap. 2), 1096.

16. James W. Girard, "Memorandum," 20 Apr. 1920, S-Studies, Logging and Milling, R-6, 54-A-111/54860, NARS-S.

17. Frank J. Klobucker to Portland Lumber Co., 18 Apr. 1921; James Girard, "Lumbering Cost Data—Team and Teamster," 20 Apr. 1921—both in S-Studies, R-1, Logging and Milling, R-6, 54-A-111/59860, NARS-S. By 1925 the two had collaborated to produce "Inland Empire: Logging Output Handbook," 1 June 1925, in the same file. In 1922 the supervisor of the Apache National Forest said the costs of sales had increased from .327 cents per MBF to .4617 cents (Silvicultural Chapter, Supervisor's Annual Plan, 1922, Apache NF, 31 Dec. 1921, in S-Plans, Annual Silvicultural Report, 1909–1929, Recreation Office, Forest Supervisor's Office).

18. Paul G. Redington (by L. W. Hess) to forest supervisors, 26 May 1922, Appraisal Methods 1913–1927, R5-Buck.

19. Press release "Douglas Fir Shows Rapid Growth," 15 Oct. 1920; E. J. Hanzlik, "Second-Growth Douglas Fir Stands Making Excellent Growth," 17 Apr. 1922, "Douglas Fir Stem Analysis Sheets," 20 Feb. 1923, and "List of Douglas Fir Stand Data," 2 May 1923; David T. Mason to Thornton T. Munger, 11 May 1923; David T. Mason and L. C. Merriam, "Douglas Fir Second Growth—What Will the Harvest Be?" 11 May 1923; E. J. Hanzlik, "Growth of Douglas Fir in Oregon," 9 Dec. 1923, a draft article published in *West Coast Lumberman,* Jan. 1924; Thornton T. Munger, "Working Plan for Study of the Yield of Douglas Fir in Western Oregon and Washington," 7 July 1924; Richard A. McArdle, "Douglas Fir Yield Study Working Plan for Office Computations," 27 Mar. 1925; Thornton T. Munger and Richard E. McArdle, "Douglas Fir: How It Grows," reprint from *West Coast Lumberman,* 1 May 1925; Walter H. Meyer, "Height Curves for Even-Aged Stands of Douglas Fir," 1 Sept. 1928; and Richard E. McArdle, "Rate of Growth of Douglas Fir Forests," reprint from *West Coast Lumberman,* 1 May 1928—all in S-Studies, Growth and Yield, Douglas Fir, R-6, 54-A-111/59860, NARS-S.

20. Robert H. Weidman, "Western Yellow Pine in the Northwest: A Study of the Silvicultural Aspects Governing Its Management (an Amplified Outline for a Bulletin)," 20 May 1921; George H. Cecil to the Forester, 28 May 1921; R. H. Weidman to DF, Portland, 4 Mar. 1930; E. J. Hanzlik, "Comments on Weidman's Report on Timber Growing, etc. of Western Yellow Pine," 24 Mar. 1930; E. L. Kolbe, "Notes on Weidman's Bulletin on Timber Growing and Logging Practices in the Northwestern Yellow Pine Region," 12 Apr. 1930; Thornton T. Munger to R. H. Weidman, 11 Apr. 1930—all in S-Studies, Publications 1913–1934, R-6, 54-A-111/59861, NARS-S; Duncan Dunning, "Relation of Crown Size and Character to Rate of Growth and Response to Cutting in Western Yellow Pine," *JF* 20 (Apr. 1922): 379–89.

21. W. B. Greeley et al., "Timber: Mine or Crop?" *Yearbook of Agriculture, 1922* (Washington, D.C.: GPO, 1923), see especially pp. 93, 103, 107, 114–17, 140–42, 150–53, 178–80.

22. "Management Plan: Crater National Forest," 23 Apr. 1921, in Rogue River NF Archive Collection, C5, Rogue River NF.

23. E. B. Birmingham, "Sustained Yield," *West Coast Lumberman,* Apr. 1937, 76, copy in S. Y. Clips-1937, PIO, PNWRO, FS.

24. Press releases, 23 and 25 Apr. and 5 Sept. 1922 and 17 Mar. 1923, USFS Clipping File, Timber Sales, FHS.

25. Press release, 1 Feb. 1923, USFS Clipping File, Timber Sales, FHS.

26. Press releases, 4 May (2 releases), 7 and 26 July 1923 and 1 Nov. 1922—all in USFS Clipping File, Timber Sales, FHS.

27. PGR [Paul G. Redington], Memorandum for Forest Management, 15 May 1923, R5-Buck; R. H. Rutledge to forest officer, 31 Dec. 1923, S-Sales, Marking 1914–1950, R-4, 60-A-192/601098, NARS-D; Lyle F. Watts, memorandum for DF [D-4], 13 Oct. 1927, S-Supervision, General, 1927–1929, R-4, 60-A-192/601106, NARS-D.

28. E. J. Hanzlik, "Some Observations on Western Yellow Pine Growth," a commentary on an article by Show, 15 Dec. 1923, S-Studies, Growth and Yield, Ponderosa Pine, R-6, 54-A-111/59860, NARS-S; J. H. Price and J. R. Berry, "White Fir Mill Scale Study," 15 June 1923, and "Yellow Pine Mill Scale Study," 4 Jan. 1924—both in Early Appraisal Accounting Methods, R5-Buck.

29. E. H. M., "Biggest Timber Sale Year," *Ranger* 2 (Apr. 1924): 9, in History—Timber Sales, Umpqua NF History Files.

30. "Management Plan Report for the Coconino National Forest," 1 Jan. 1923, and Sherman to DF, 7 Mar. 1923—both in TSO files, Coconino NF.

31. Julius F. Kummel, "Methods and Season of Direct Seeding, Western Yellow Pine," 19 Sept. 1924, S-Studies, Season and Method of Sowing and Planting, R-6, 54-A-111/59862, NARS-S; W. W. White, "Growth after Cutting on a Western Yellow Pine Timber Sale Area in Montana," Region 1 *Applied Forestry Notes,* no. 48 (1 June 1924); J. O. Thompson, "Yellow Pine Silvicultural Practice on the Custer National Forest," *Applied Forestry Notes,* no. 47 (1 May 1924).

32. E. J. Hanzlik, "Growth of Reserve Trees on Western Yellow Pine Cut-Over Areas in Oregon," 16 Jan. 1925, and appended memorandums, S-Studies, Growth and Yield, Ponderosa Pine, R-6, 54-A-111/59860, NARS-S.

33. W. W. Ashe, "Economic Waste in Cutting Small Timber," *Southern Lumberman,* 20 Dec. 1924, 184–187.

34. Paul G. Redington to [eight timber companies], 27 Mar. 1924; J. C. Elliott, Memorandum for Mr. Woodbury, 12 Nov. 1924; and L. F. Kellogg, " 'Cat'-Wheel Logging," 9 Mar. 1925—all in Appraisal Methods, 1913–1927, R5-Buck.

35. *Statistical History of the United States,* 541; Edward M. Davis, *Industrial Outlets for Short-Length Softwood Yard Lumber,* USDA circ. 393, July 1926 (Washington, D.C.: GPO, 1926).

36. Marc W. Edwards, "Logging—Damage Study," 8 Jan. 1926, and T. D. Woodbury to forest supervisors, 14 Jan. 1926, Appraisal Methods, 1913–1927, R5-Buck. The company in question was Fruit Growers Supply, discussed above.

37. Press release, 5 Apr. 1927, USFS Clipping File, Timber Sales, FHS.

38. Edward A. Sherman, *Forestry as a Profession* (Washington, D.C.: GPO, 1927), 1; news release, National Lumber Manufacturers Association (NLMA) (8 June 1927), box 123, Sustained Yield, NLMA Records, FHS.

39. Management Plan, Coconino NF, approved 26 July 1927, Management Plans—Coconino Flagstaff Working Circle, TSO files, Coconino NF.

40. McNary-Woodruff Act, 45 Stat. 468; McSweeney-McNary Act, 45 Stat. 699; Steen, *U.S. Forest Service,* 193–94.

41. Steen, *U.S. Forest Service,* 197–98.

42. McGeary, *Gifford Pinchot,* 330–35.

43. C. N. Woods, Memorandum for Forest Management, 5 Dec. 1928, S-Supervision, General, 1927–1929, R-4, 60-A-192/601106, NARS-D.

44. Steen, *U.S. Forest Service,* 251; Elmo Richardson, *David T. Mason: Forestry Advocate* (Santa Cruz, Calif.: FHS, 1983).

45. Carl M. Stevens to R. H. Rutledge, 9 Jan. 1928, S-Supervision, General, 1927–1929, R-4, 60-A-192/601106, NARS-D.

46. Mason to Rutledge, 16 Jan. 1928, S-Supervision, General, 1927–1929, R-4, 60-A-192/601106, NARS-D.

47. Rutledge to Stevens, 21 June 1928, S-Supervision, General, 1927–1929, R-4, 60-A-192/601106, NARS-D.

48. USDA, FS, Calif. Dist., *Forest Management Handbook* (mimeo, looseleaf, 1928; 114 pp.), 2. The copy I examined was in the TMO, Sequoia NF.

49. Western Forestry and Conservation Association, *Cooperative Forest Study of the Grays Harbor Area* (Portland, Oreg.: Western Forestry and Conservation Association, 1929); Edwin Van Syckle, *They Tried to Cut It All* (Aberdeen, Wash.: Friends of the Aberdeen Public Library, 1980), especially at 294–95. Breakage and other waste from careless logging had occupied Forest Service researchers for several years: see, for examples, E. F. Rapraeger, ''Tree Breakage and Felling Practices,'' Aug. 1932, S-Studies, Tree Breakage and Felling Practice in Douglas Fir, R-6, 54-A-111/59862; E. J. Hanzlik and F. S. Fuller, ''A Study of Breakage, Defect and Waste in Douglas Fir,'' *Washington University Forest Club Annual* 5 (1917): 32–39; Allen H. Hodgson, ''Logging Waste in the Douglas Fir Region,'' published serially in 1929 and reprinted in Jan. 1930 by *The Pacific Pulp and Paper Industry* and *West Coast Lumberman;* J. S. Boyce et al., *Losses in Douglas Fir in Western Oregon and Washington,* USDA Agricultural Bulletin 286 (Washington, D.C.: GPO, 1932); A. H. Hodgson, ''To Reduce Waste of an Important Natural Resource,'' 23 Sept. 1932, S-Studies, Forestry Program for Oregon and Washington, 1932, R-6, 54-A-111/59859, NARS-S.

50. G. A. Pearson, *Management of Ponderosa Pine in the Southwest,* Agriculture Monograph no. 6 (Washington, D.C.: GPO, 1950), 47.

51. Ronald B. Hartzer, *Half a Century in Forest Conservation: A Biography and Oral History of Edward P. Cliff,* Report DAC-4, prepared under contract for the USDA, FS, David A. Clary, project director (Bloomington, Ind.: David A. Clary & Associates, 1981), 247–48.

52. ELK [Kolbe], ''Growth after Cutting of Western Yellow Pine in Three Localities of Southern Oregon,'' Mar. 1929, transmitted with Thornton T. Munger to DF, 30 Apr. 1929, S-Studies, ME, Ponderosa Pine, R-6, 54-A-111/59860, NARS-S.

53. See George L. Drake to director, Pacific Northwest Forest Experiment Station, 7 Dec. 1929; Thornton T. Munger to DF, Portland, 24 Dec. 1929; and Munger to RF, Portland, 16 July 1930—all in S-Studies, ME, Ponderosa Pine, R-6, 54-A-111/59860, NARS-S; William H. Gibbons, Herman M. Johnson, and Howard Spelman, ''The Effect of Tree Sizes on Western Yellow Pine Lumber Values and Production Costs,'' in six parts, *Timberman* 30 and 31 (1929 and 1930);

I. V. Anderson, *Application of Selective Logging to a Ponderosa Pine Operation in Western Montana* (Missoula: State University of Montana, n.d. [1930?]).

54. Axel Brandstrom, "Preliminary Report on Franklin River Study," 30 Aug. 1930, S-Studies, Logging Studies—Selective—Brandstrom, R-6, 54-A-111/59861, NARS-S; Greeley, introduction to Axel J. F. Brandstrom, *Analysis of Logging Costs and Operating Methods in the Douglas Fir Region* (published by the Charles Lathrop Pack Forestry Foundation under the auspices of the West Coast Lumbermen's Association, June 1933), 2–3.

55. Walter H. Meyer, "A Method of Constructing Yield Tables for Selectively Cut Stands of Western Yellow Pine," 15 Mar. 1930, an article to be submitted to *JF*, S-Studies, ME, 101-Selectively Cut Stands (Ponderosa Pine), R-6, 54-A-111/59860, NARS-S.

56. Roland Rotty to RF, 20 Jan. 1949, in TMO files, 2410 Plans, SWRO, USFS.

57. H. H. Chapman, *Forest Management* (New York: John Wiley & Sons, 1931), 79; W. C. [Wilson Compton] to E. W. Demarest, 12 Apr. 1939, NLMA Records, box 21, "Sustained Yield Forestry Legislation," FHS. There was little demand for lumber during the depression. Construction in the United States fell from $13.9 billion in 1929 to $5.7 billion in 1932 (Clary, *Life Which Is Gregarious*, 199).

58. Chapman, *Forest Management* (1931), 78.

59. USDA, FS, Dist. 6, *Instructions for Preparation of the Nineteen Thirty Timber Inventory of the National Forests: Douglas Fir Region, District Six* (Portland, Oreg.: USDA, FS, 1930).

60. "Report of Requirements Committee, Forest Survey, Madison Conference," May 1930, and "Report of Growth and Drain Committee, Forest Survey, Madison Conference," May 1930, S-Studies, Forest Survey, R-6, 54-A-111/59859, NARS-S.

61. James W. Girard, *The Man Who Knew Trees: The Autobiography of James W. Girard* (St. Paul: Forest Products History Foundation and Minnesota Historical Society, 1949), 25–28; Alfred Wiener, personal communication.

62. USDA, FS, *Facts Concerning the Forest Survey, Its Scope and Value* (Washington, D.C.: USDA, FS, 1930).

63. "Policy Statement, Lincoln National Forest," May 1931, file 1680, Lincoln NF, "Policy Statement, Crook National Forest," 30 June 1931, and D. M. Lang to forest supervisor, 29 Sept. 1936, History File, Apache NF.

64. Axel J. F. Brandstrom, "An Estimate of the Economic Effect of Selective Logging in the Douglas Fir Region," 10 Nov. 1931, S-Studies, Logging Studies, Selective, Brandstrom, R-6, 54-A-111/59861, NARS-S; Hessler & Co., "Working Plan: Mill Scale Studies, Douglas Fir and Hemlock," June 1931; Allen H. Hodgson, "Summaries of Individual Sawmill Waste Studies Conducted in Four Hemlock and Six Douglas Fir Sawmills of Western Washington and Oregon," 29 Jan. 1931; and I. V. Anderson, "Residual Wood after Logging in the Western White Pine Region," Jan. 1932—the last three in S-Studies, Logging and Milling, R-6, 54-A-111/59860, NARS-S.

65. Ovid Butler to S. B. Detwiler, Bureau of Plant Industry, S Disease Control, 1928–37, Records of the FS, RG95, box 1339, NA, Washington, D.C. (hereafter Forest Service records in the National Archives will be cited RG95/box number); see also Warren V. Benedict, *History of White Pine Blister Rust Control: A Personal Account* (Washington, D.C.: USDA, FS, 1981).

66. *A National Plan for American Forestry,* Sen. Doc. 12, 73d Cong., 1st sess. (1933); Steen, *U.S. Forest Service,* 199–204. Correspondence during preparation of the report is in S-Studies, Copeland Resolution, R-6, 54-A-111/59858, NARS-S.

CHAPTER FOUR. TIMBER MANAGEMENT TAKES CONTROL

1. EHC [Clapp] to Buck, 9 May 1933, S-Studies, Copeland Resolution Correspondence, 1932–1935, R-6, 54-A-111/59858.

2. Steen, *U.S. Forest Service,* 327. The background of that action is presented by Horace M. Albright in *Origins of National Park Service Administration of Historic Sites* (Philadelphia: Eastern National Park & Monument Association, 1971) and by Ronald F. Lee in *The Antiquities Act of 1906* (Washington, D.C.: NPS, 1970).

3. The standard history of the CCC is John A. Salmond's *The Civilian Conservation Corps, 1933–1942* (Durham, N.C.: Duke University Press, 1967); see also Steen, *U.S. Forest Service,* 213–16; Pyne, *Fire in America,* 360–70; and Charles Lathrop Pack, *White Pine Blister Rust: A Half-Billion Dollar Menace* (Washington, D.C.: Charles Lathrop Pack Forestry Foundation, 1933), 1. The equipping of the CCC and the construction of its camps are discussed by Clary in *Life Which Is Gregarious,* 200–203; see also Francis V. FitzGerald, ''The President Prescribes: The Organization and Supply of the Civilian Conservation Corps,'' *Quartermaster Review* 13 (July/Aug. 1933): 7–18; John A. Porter, ''The Enchanted Forest: An Account of the Civilian Conservation Corps,'' *Quartermaster Review* 13 (Mar./Apr. 1934): 5–17; and Enoch Graf, ''The Army's Greatest Peace-Time Achievement,'' *Quartermaster Review* 15 (July/Aug. 1936): 15–20.

4. National Industrial Recovery Act, 48 Stat. 195; Steen, *U.S. Forest Service,* 226–27; Arthur M. Schlesinger, Jr., *The Coming of the New Deal* (Boston: Houghton Mifflin, 1958, 1965), 87–176. The case that vacated NIRA was *Schecter* v. *United States,* 295 U.S. 495 (1935), which related to the wholesaling of chickens in New York. On the origins of the code see ''Conference of Lumber and Timber Products Industries with Public Agencies on Forest Conservation,'' *JF* 32 (Mar. 1934): 275–307; and especially the outstanding history in Richardson, *David T. Mason,* 41–68.

5. Steen, *U.S. Forest Service,* 227.

6. William B. Greeley, ''Industrial Forest Management in the Pacific Northwest as Influenced by Public Policies,'' Duke University School of Forestry Lectures no. 7 (Apr. 1948): 9–10.

7. Natural Resources Board, *A Report on National Planning and Public Works in Relation to Natural Resources and Including Land Use and Water Resources with Findings and Recommendations* (Washington, D.C.: GPO, 1934), esp. 2, 4, 21, 206, 209. Steen, *U.S. Forest Service,* 227–44, discusses the renewed wrangles over regulation and the transfer of the national forests to Interior. For renewed efforts to implement the Copeland Report see Clapp to RF, 12 Apr. 1934, and Secretary of Agriculture Henry A. Wallace to the president, 18 Apr. 1934, S-Legislation, 1920–1934, Closed, R-6, 54-A-111/59854, NARS-S.

8. G. F. Jewett, ''An American Plan for Forestry,'' *JF* 32 (Mar. 1934): 308–12, which called for dividing the entire country into ''logical sustained yield units''; *Astorian Budget* (Oreg.), 18 Dec. 1934, ''Buck Shows Need of Sustained Yield of Forest Products''; and similar articles in at least twelve other newspapers in Oregon, Washington, and Idaho, in Dec. 1934—clippings in History Collection,

Sustained Yield Clippings 1934–35, PIO, PNWRO, FS. On Forest Survey findings see Herman M. Johnson, "Cutting Depletion Phase of the Forest Survey in Western Washington and Western Oregon," Jan. 1934; Thornton T. Munger to RF, 15 June 1934; F. H. Brundage to Munger, 21 June 1934; and Brundage to Munger, 22 Oct. 1934—all in S-Studies, FS, Depletion Phase, R-6, 54-A-111/59862, NARS-S; Munger to RF, 30 Jan. 1934; F. V. Horton to Munger, 7 Feb. 1934; Fred Ames to Munger, 8 Feb. 1934; Munger to RF, 14 Feb. 1934; F. H. Brundage for C. J. Buck to director, Pacific Northwest Forest Experiment Station, 21 Feb. 1934; and C. Weldon Kline, "Comparisons of National Forest Timber Estimates in the Douglas Fir Region," 20 Dec. 1934—all in S-Studies, FS, Douglas Fir, R-6, 54-A-111/59859, NARS-S.

9. F. H. Brundage for C. J. Buck to director, Pacific Northwest Forest Experiment Station, cited in note 8; Fred Merkle, "Management Plan, Flagstaff Working Circle—Decade 1933–1942," approved 16 Feb. 1934; E. E. Carter, "Memorandum for Files," 26 Feb. 1934; and L. K. Kneip to RF, 26 Feb. 1934—all in TSO files, Coconino NF.

10. "Digest Report" of Regional Foresters and Directors Annual Meeting, 24 Nov. to 8 Dec. 1936, D-Supervision, Meetings, 1936, R-6, 54-A-117/21127, NARS-S.

11. F. A. Silcox, "Wildlife Management and the National Forests," address before the Twenty-first Annual Game Conference at New York City, 23 Jan. 1935, NLMA Records, box 54, U.S. Government—USFS, FHS.

12. *Register-Guard* (Eugene, Oreg.), 28 Feb. 1935, "Buck Reveals Forest Plans of Vast Area," in Sustained Yield Clippings 1934–35, PIO files, PNWRO, FS.

13. *East Oregonian* (Pendleton), 16 Feb. 1935, and contemporary clippings from newspapers of Portland, Grants Pass, Klamath, Union County, Wasco County, Douglas County, Jackson County, Clatsop, Bend, and others—all in Sustained Yield Clippings 1934–35, PIO files, PNWRO, FS.

14. *Oregon Journal* (Portland), 10 Feb. 1935, "Timber Yield Program for County Asked," in Sustained Yield Clippings 1934–35, PIO files, PNWRO, FS.

15. *Chronicle* (Elma, Wash.), 21 Feb. 1935, "Sustained Timber Yield Is Visioned," in Sustained Yield Clippings 1934–35, PIO files, PNWRO, FS.

16. There is an extremely thick collection of clippings and releases documenting every step of the controversy in the PIO files, PNWRO, FS. Edward P. Cliff, who was the regional wildlife officer of the Forest Service at the time, explains the wildlife dimension of the dispute in Hartzer, *Half a Century*, 91–93; see also Alfred Runte, *National Parks: The American Experience* (Lincoln: University of Nebraska Press, 1979), 141–42. I am also grateful for the account of Roosevelt's trip in a personal communication from Glen Jorgensen.

17. "Proposed Program of Study in the Chemical Control of Vegetation," 13 Sept. 1934, in Chemical Control of Vegetation, file 67-1099/224013, Forest Fire Laboratory, Laguna Niguel; USDA, FS, *Instructions and Information Concerning Stand Improvement Work, California Region,* 23 Mar. 1934, and *Annual Stand Improvement Report, 1934, Region 5,* 4 Feb. 1935—both in TMO files, Sequoia NF; USDA, FS, Division of Silvics, *Sample Plots in Silvicultural Research,* USDA circ. no. 333 (Washington, D.C.: GPO, 1935); "A Survey of Forest Insect Infestations on Cutover Areas in California," 1935, FS-Experiment Station, Correspondence, 1924–1944, 60-A-112/69278, NARS-San Bruno, Calif. (NARS-SB); Philip Neff, "Cost of Trucking White Pine Logs," 6 Mar. 1936, S-Supervision, General, 1936–37, R-4, 60-A-192/601106, NARS-D; Thornton T. Munger to RF, 11 May

1935, and Donald N. Mathews, "Memorandum to Accompany the Fire Depletion Statistics of the Douglas Fir Region, Prepared for the Forest Survey," 15 Jan. 1935, S-Studies, FS, Depletion Phase, R-6, 54-A-111/59862, NARS-S.

18. F. P. Keen, "Ponderosa Pine Tree Classes Redefined," *JF* 34 (Oct. 1936): 919–27; J. Elton Lodewick to RF, 15 Jan. 1936; J. Elton Lodewick and Herman M. Johnson, Memorandum, 24 Jan. 1936; H. J. Andrews to RF, 3 May 1935; and J. Elton Lodewick, "Pulpwood Conversion Factors in the Pacific Northwest," Apr. 1935—all in S-Studies, Logging and Milling, 1918–1936, R-6, 54-A-111/59860, NARS-S; *Times-Herald* (Burns, Oreg.), 28 July 1936, "Lighter Cut Trend Talked," and *Independent* (Helena, Mont.), 16 Sept. 1937, "Example in Forestry"—both in Sustained Yield Clippings, 1935–36, PIO files, PNWRO, FS; I. V. Anderson, "Why Selective Cutting of Ponderosa Pine Pays," *Timberman* 35 (Sept. 1934): 3–12; Thornton T. Munger, Axel J. F. Brandstrom, and Ernest L. Kolbe, "Basic Considerations in the Management of Ponderosa Pine Forests by the Maturity Selection System," PNW Research Paper, 1 Sept. 1936; and E. E. Carter to RF, 18 June 1936—both in S-Sales, Marking, 1914–1950, R-4, 60-A-192/601098, NARS-D. In 1936, Raphael Zon told the regional foresters that in the Soviet Union, "clear-cutting is absolutely taboo and practically all operations are on a light selective basis" ("Digest Report" of RF&D annual meeting, 24 Nov. to 8 Dec. 1936, D-Supervision, Meetings, 1936, R-6, 54-A-117/21127, NARS-S).

19. *Sun* (Hood River, Oreg.), 30 Sept. 1936, "Timber!"; *Lake County Tribune* (Lakeview, Oreg.), 17 Sept. 1936, "M'Nary Sees Conservation Need Locally"; *Mail* (Myrtle Creek, Oreg.), 13 Aug. 1936, "Suggests Change Forest Handling"; *Journal* (Portland, Oreg.), 9 Aug. 1936, "Forest Experts Urge Scientific Planning"; *Evening News* (Port Angeles, Wash.), 4 Dec. 1936, "Sustained Yield Forestry Bill Gets Approval"; *Tidings* (Ashland, Oreg.), 24 Nov. 1936, "Conservation Lets Us Eat and Have Our Cake"; *Sno County Forum* (Granite Falls, Wash.), 11 June 1936, "Sustained Yield, Selective Logging to Be Developed"—all in Sustained Yield Clippings, 1935–36, PIO files, PNWRO, FS.

20. "Digest Report," annual meeting of regional foresters and directors, 24 Nov. to 8 Dec. 1936, D-Supervision, Meetings, 1936, R-6, 54-A-117/21127, NARS-S.

21. Ibid.

22. The following summary of the regulation controversy is based on Steen, *U.S. Forest Service,* 229–37. Emanuel Fritz of the University of California remarked: "My belief is that there are several men next to Silcox who do not want private effort to succeed and will sabotage any cooperative plans favored by their more conservative colleagues" (Fritz to G. F. Jewett, 31 Jan. 1938, NLMA files, box 123, Sustained Yield, FHS. The NLMA records on the controversy are abundant. Regarding Greeley's position on regulation (below) see in particular his comments on the report of the Joint Committee on Forestry, 10 Apr. 1941, in box 25, Congressional Joint Committee.

23. Steen, *U.S. Forest Service,* 234–35.

24. Peck to Regional Forester Rutledge, 18 Dec. 1937, D-Supervision, Meetings, 1938, R-6, 54-A-117/21127, NARS-S.

25. E. E. Carter to RF, 19 and 24 Sept. 1937; and R. H. Rutledge to chief, 23 Nov. 1937, S-Supervision, General, 1936–1937, R-4, 60-A-192/601106, NARS-D. Two foresters were ordered to write a new one in 1937.

26. "Re-seeding Cut Over Lands," *West Coast Lumberman,* May 1937, 22; Axel J. Brandstrom, Clarence C. Richen, and Ray C. Carlson, "Memoranda

Covering the Analysis of the Experimental Light Marking on the Edw. Hines Lumber Co. Timber Sale,'' 12, 13, and 19 Feb. 1937, and Axel J. F. Brandstrom, ''Memorandum Re: Comments on the New Marking Rules for the Hines Sale,'' 11 Aug. 1937, conveying memoranda of 3 May, 25 June, and 11 Aug. 1937—both in S-Studies, Logging and Milling, Edw Hines Lbr Co., R-6, 54-A-111/59860, NARS-S; Buck to John Boettiger, 3 Nov. 1937, D-Supervision, President's Trip, in R-6, 1937, R-6, 54-A-117/21127, NARS-S.

27. *Bulletin* (Bend, Oreg.), 9 Oct. 1923, ''Sustained Yield Problem Discussed at Meeting Here''; and *Daily World* (Wenatchee, Wash.), 1 Dec. 1937, ''U.S. Starts Tree Trades''—both in Sustained Yield Clippings 1937, PIO files, PNWRO, FS; Benson H. Paul, *Knots in Second-Growth Pine and the Desirability of Pruning*, USDA Miscellaneous Publication no. 307 (Washington, D.C.: GPO, 1938); Fred R. Mason, ''Hines Lumber Company Operation at Burns, Oregon,'' 7 July 1939, S-Supervision, General, 1938–1942, R-4, 60-A-192/601106, NARS-D.

28. USDA, FS, *Coconino National Forest, Arizona* (Washington, D.C.: GPO, 1939), 9–10; Tucker, ''Forest Service in the Southwest,'' 1263.

29. See the following in Sustained Yield Clippings, 1938–39, 1940, 1941, PIO files, PNWRO, FS: *Oregon Journal* (Portland), 13 Feb. 1938, ''Forest Man's Warning Spurs Extended Quiz''; *Valley Ranger* (John Day, Oreg.), 26 Aug. 1938, ''Pine and Fir Lumbermen Meet with Planning Board''; *Columbian* (Vancouver, Wash.), 16 Sept. 1938; *Sunday Ledger and News Tribune* (Tacoma, Wash.), 9 Oct. 1938, ''Timbermen Meet Soon''; *Timber Worker* (Seattle, Wash.), 28 Jan. 1939, ''Ghost Towns Seen If Action Not Taken; 2 Billion Industry Can Be 5 Billion, If Regulated''; *Daily News* (Eugene, Oreg.), 28 May 1939, ''Markets Are Crying Need of Lumber Industry Now, Declares J. H. Cox''; *Oregon Journal* (Portland), 12 July 1939, ''Many Sawmills Are Too Large, Silcox Declares''; *Bulletin* (Bend, Oreg.), 15 July 1939, ''Forest Yield Problems Are Seen on Tour,'' and 8 Aug. 1939, ''Sustained Yield Plan Is Adopted''; *Post Intelligencer* (Seattle, Wash.), 30 Sept. 1939, ''Wanted: A Forest Policy''; *Christian Science Monitor,* 7 Dec. 1939, ''A Window on the West''; *Union Register* (Seattle), 15 Dec. 1939, ''Conservation Being Used by IWA to Cover Up Fast Falling Morale''; *Post Intelligencer* (Seattle), Dec. 1939, ''A Forest Program Still Awaited''; ''Shevlin-Hixon Launches Sustained Yield Plan,'' *West Coast Lumberman,* Jan. 1940; *Post Intelligencer* (Seattle), 18 Jan. 1940, ''Wanted—A Program''; *Evening News* (Port Angeles, Wash.), 31 Jan. 1940; *Post Intelligencer* (Seattle), 20 Feb. 1940, ''Let's Have That Forest Program''; *Register-Guard* (Eugene, Oreg.), 27 Feb. 1940, ''Harvesting Our Timber Crops''; *Spokesman* (Redmond, Oreg.), 21 Mar. 1940, ''For the Small Fellow, a Kick in the Teeth,'' and 25 Apr. 1940, ''Who Does Know the Answer?''; *Sentinel* (Goldendale, Wash.), 11 Apr. 1940, ''This Is Conservation Era''; and *Vidette* (Montesano, Wash.), 10 June 1941, ''Great Clemons Tree Farm Heralded.''

30. Keen, ''Ponderosa Pine Tree Classes,'' 10 Feb. 1941; Lyle F. Watts to forest officer, 16 July 1941; F. V. Horton to RF, Region 4, 10 Nov. 1941; W. L. Robb to forest supervisor, 21 Feb. 1942; and Robb to F. P. Keen, 17 Mar. 1942—all in S-Sales, Marking, 1914–1950, R-4, 60-A-192/601098, NARS-D; ''More Snow in Thinned Forests,'' 1 Mar. 1940, S-Studies, Alaska Pulp and Paper, R-6, 54-A-111/59862, NARS-S; Stephen N. Wycoff to F. C. Craig, 5 Sept. 1941, FS-Experiment Station, Correspondence, 1924–1944, 60-A-112/69278, NARS-SB; ''Forest Insects,'' 21 Feb. 1940, in Archeological/Historical Collection, Shasta Forest Historical Insect Control, Shasta-Trinity NF; James E. Sowder and Russel W. Beeson, ''Management Plan for the Big Valley Working Circle, Modoc

National Forest,'' 9 Apr. 1940, in TMO files, file 2410, Plans, Big Valley Federal Sustained Yield Unit, Modoc NF.

31. Minutes of Regional Foresters Conference, 8 Feb. 1940, in J. H. Stone Collection, Articles, Speeches and Statements, J. H. Stone: 1936–1950, PNWRO, FS.

32. Earle H. Clapp to RFs and directors, 30 Mar. 1940; and C. L. Billings, '' 'Canned Heat,' or Forest Regulation by the National Government by Fair Means—or Otherwise,'' speech, 6 June 1941—both in NLMA files, box 22, Forest Conservation Committee Meeting, exhibits, 10 and 11 Nov. 1941, FHS.

33. *Morning Herald* (Yakima, Wash.), 15 Dec. 1940, ''Lumber Industry Geared to Speed,'' in Sustained Yield Clippings, 1940, PIO files, PNWRO, FS.

34. I discuss mobilization construction in *Life Which Is Gregarious,* 208–301.

35. *Statistical History of the United States,* 534, 541. There is a built-in statistical error in comparing figures for timber production and lumber production, because they are based on different meashing techniques. Generally, one MBF of timber (log scale) yields more than one MBF of lumber (mill tally).

36. *Spokesman* (Redmond, Oreg.), 21 Mar. 1940, ''For the Small Fellow, a Kick in the Teeth,'' in Sustained Yield Clippings 1940, PIO files, PNWRO, FS.

37. Steen, *U.S. Forest Service,* 246–53, summarizes the wartime period; see also Girard, *Man Who Knew Trees,* 30–32.

38. Pyne, *Fire in America,* 370–73; USDA, FS, *The Use of Prisoners of War in Logging, Pulpwood, and Lumber Industries* (n.p., n.d. [c. 1943/44]), available at FHS; Wellington R. Burt to A. G. T. Moore, 24 Feb. 1944, USDA news release ''U.S. Salvage Program for Storm-Damaged Timber,'' 20 Feb. 1944, and P. F. Hursey to Erle Kauffman, 11 Feb. 1944—all in NLMA files, box 22, Texas Timber Salvage; Report of General Inspection of the Inyo National Forest, 1945, D-Inspections, J. E. Elliott, Frank J. Jefferson, 8/24-9/20/45, 64-300/44017, NARS-Laguna Niguel (NARS-LN); *News* (Hood River, Oreg.), 27 Aug. 1943, ''And Selective Logging, Too,'' News Clippings, Selective Logging, 1939–43, Sustained Yield, 1941–46, PIO files, PNWRO, FS; see also Edward P. Cliff's wartime reminiscences in Hartzer, *Half a Century,* 131–38.

39. Robert E. Ficken, ''Pulp and Timber: Rayonier's Timber Acquisition Program on the Olympic Peninsula, 1937–1952,'' *JFH* 27 (Jan. 1983): 4–14.

40. Carter to RF, San Francisco, 4 July 1945, DTMRF RG95/1369.

41. Carter to RF, Portland, 1 Aug. 1945, DTMRF, RG95/1369.

42. The regulation controversy that finally ended in 1952 was an exceedingly complex story, most of which is beyond the scope of this study. The summary here follows Steen, *U.S. Forest Service,* 259–71.

43. Ibid., 256–58; John B. Woods, ''Report of the Forest Resource Appraisal,'' *American Forests* 52 (Sept. 1946): 414–28.

44. Guy Cordon, ''Uncle Sam, Land Owner [guest editorial],'' in *Journal* (Portland, Oreg.), 31 Jan. 1946, News Clippings, 1945–1949, PIO files, PNWRO, FS.

45. *Journal* (Portland, Oreg.), 31 Jan. 1946, ''Federal Tree Hoarding Hit''; and *News-Review* (Roseburg, Oreg.), 11 Jan. 1946, News Clippings, 1945–1949, PIO files, PNWRO, FS.

46. Specimens of the hoarding controversy in the press are the following from News Clippings, 1939–1943, and News Clippings, 1945–1949, PIO files, PNWRO, FS: *News-Review* (Roseburg, Oreg.), 25 Jan. 1946, ''Says Demand for Housing May Halt Sustained Yield''; ''Hoarded Timber and Other Matters,'' *International Woodworker,* May 1946; *Times* (Seattle), 17 May 1946, ''Opening of

U.S. Timber Lauded''; *Register-Guard* (Eugene, Oreg.), 19 May 1946, ''Lumberman Calls for Honest Federal Timber Management''; *Oregon Journal* (Portland), 16 June 1946, ''Lumberman, Forest Service Odds May Slow Vet Housing''; Lamar Newkirk, ''Northwest Woods,'' in *Oregon Journal* (Portland), 28 July 1946; *Oregonian* (Portland), 10 Sept. 1946, Burt P. Kirkland, ''To the Editor,'' 18 Aug. 1946, ''Forest Yield Gains Seen,'' and 31 Aug. 1946, ''The Kirkland Report.'' The contrivance of the issue is suggested by the fact that the West Coast Lumbermen's Association had claimed just a year earlier that ''there is no danger of the forests being completely cut away as long as the sustained yield program is followed with our trees'' (*Daily World* [Wenatchee, Wash.], 23 Mar. 1945, ''Plenty Lumber, Now and for Posterity). On the NLMA statement see Steen, *U.S. Forest Service,* 258.

47. Carter to RF, Portland, 1 Aug. 1945, DTMRF, RG95/1369.

48. Carter to the files, 28 Aug. 1945, DTMRF, RG95/1369.

49. Carter to the files, 29 Oct. 1945, DTMRF, RG95/1368.

50. Mason to RFs, 24 Sept. 1945, DTMRF, RG95/1369.

51. Mason to Irving Cheskin, National Housing Agency, 13 Mar. 1946; Granger to Richard M. Bissell, Jr., Office of War Mobilization and Reconversion, 4 Mar. 1946; and Mason, ''Notes on Statement of Proposed F. Y. 1947 FRD Road Program,'' 15 Mar. 1946—all in DTMRF, RG95/1368.

52. Mason to Raymond Marsh, 23 July 1946, DTMRF, RG95/1368.

53. Mason to the files, 1 Aug. 1946; C. M. Granger to RFs, 9 Sept. 1946; and H. E. Ochsner to Granger, 23 Sept. 1946, DTMRF, RG95/1368.

54. Mason to Messrs. Granger and Jones, 7 Nov. 1947, DTMRF, RG95/1367.

55. Mason to Division of Operations, 23 Apr. 1948, DTMRF, RG95/1336; *San Francisco Chronicle,* 28 Aug. 1950, ''More Funds for Logging Roads Urged,'' NLMA files, box 54, U.S. Govt.—USFS, FHS. The problem ultimately was solved in part by crediting timber purchasers for permanent roads that they built.

56. ''Management Plan, Flagstaff Working Circle, Decade 1943–1952,'' 12 Mar. 1943; ''Supplement to Timber Management Plan, Flagstaff Working Circle, Decade 1943–1952,'' 12 Aug. 1947; and Mason to the files, 26 Sept. 1947—all in TSO files, Coconino NF.

57. Watts to secretary of agriculture, 22 Mar. 1945; and Mason to RF, Philadelphia, 12 May 1945, DTMRF, RG95/1369.

58. Carter to C. M. Granger, 16 Apr. 1945, DTMRF, RG95/1369.

59. Lockmann, *Guarding the Forests of Southern California;* Tweed, *Recreation Site Planning;* C. M. Granger to James W. Girard, 19 May 1945, DTMRF, RG95/1369.

60. Mason to Carter, 12 May 1945; and Carter to RF, San Francisco, 4 July 1945—both in DTMRF, RG95/1369.

61. Mason to C. M. Granger, 24 May 1945; Mason to the files, 1 and 25 June 1945; and A. W. Greeley to Carter, 17 July 1945—all in DTMRF, RG95/1369.

62. Mason to RF, Missoula, 31 Oct. 1945; Carter to RF, Portland, 1 Aug. 1945; and Mason to C. M. Granger, 20 Sept. 1945—all in DTMRF, RG95/1368 and 1369.

63. Division of Timber Management 1945 Accomplishment Statement, n.d. [11 Dec. 1945]; and Mason to Granger, 15 Feb. 1946—both in DTMRF, RG95/1368.

64. Mason to H. D. Cochran, Division of Personnel Management, 22 Mar. 1946; Mason to RFs and Tropical Forestry Unit, 7 Mar. 1946; and Mason to Dana

Parkinson, Division of Information and Education, 22 Mar. 1946—all in DTMRF, RG95/1368.

65. Leon Kneipp to RFs, 11 Apr. 1946; Mason to L. S. Gross, 30 Apr. 1946; and Watts to John Boettiger, 31 May 1946, DTMRF, RG95/1368.

66. H. E. Ochsner to Mason, 27 June 1946; and Mason to RF, Ogden, 27 Mar. 1946—both in DTMRF, RG95/1368.

67. Mason to R. E. Marsh, 23 July 1946, DTMRF, RG95/1368.

68. "Report to the Select Committee on Newsprint and Paper Supply," 23 July 1947; Watts to Ralph R. Will, in re Bill S. 788, 24 Mar. 1947; and L. S. Gross to Mason, 12 May 1947—all in DTMRF, RG95/1367.

69. C. M. Granger to R-8, 6 Oct. 1947; and Mason to Granger, 12 Oct. 1948—both in DTMRF, RG95/1367 and 1366. On the region's grazing controversies see Hartzer, *Half a Century,* 11n, 12–14, 166–78; Steen, *U.S. Forest Service,* passim; and William D. Rowley, *U.S. Forest Service Grazing and Rangelands: A History* (College Station: Texas A & M Press, 1985).

70. Mason to R-6, 2 Jan. 1948, DTMRF, RG95/1367.

71. Mason to Division of Operations, 23 Apr. 1948, DTMRF, RG95/1366.

72. Granger to Flannagan, 3 May 1948, to the editor, 24 May 1948, and to RFs, Nov. 1948—all in DTMRF, RG95/1366.

73. Payne to the files, 13 May 1948; Gross to R-4, 12 May 1948; Gross to the record, 25 Aug. 1948—all in DTMRF, RG95/1366.

74. "Timber Management Plan, North Kaibab Working Circle, Kaibab National Forest, Arizona," Mar. 1948, in Supervisors Office files, 2410 Plans, Kaibab NF; *Trees: The Yearbook of Agriculture 1949* (Washington, D.C.: GPO, 1949), *San Francisco Chronicle,* 28 Aug. 1950, "More Funds for Logging Roads Urged," NLMA files, box 54, FHS.

75. J. H. Stone, "The Federal View of Forestry Legislation," address to Southern Forestry Conference, 17–18 Feb. 1950, in Articles, Speeches and Statements, J. H. Stone, 1936–1950, J. H. Stone Collection, PNWRO, FS; Region 6 Timber Management Handbook, pt. 5, 1950, copy in files of Rogue River National Forest; Pearson, *Management of Ponderosa Pine in the Southwest,* vi, 13.

76. *Oregonian* (Portland), 15 Aug. 1950, "Forester Acts on War Peril"; "Battle of the Beetles," Edward P. Cliff Papers, vol. 1, FS History Section, 7 June 1951; S. E. Jarvi to the files, 14, 15, and 20 Aug. 1952, and William A. Peterson to supervisors, 6 Oct. 1952—both in Chemical Control of Vegetation, Forest Fire Laboratory, 67/1099/224013, NARS-LN.

77. *Statistical History of the United States,* 534.

CHAPTER FIVE. ADVENTURES IN LEGISLATIVE SUSTAINED YIELD

1. Sustained-Yield Forest Management Act of 1944, 58 Stat. 132; Steen, *U.S. Forest Service,* 251–52; Roy O. Hoover, "Public Law 273 Comes to Shelton: Implementing the Sustained-Yield Forest Management Act of 1944," *JFH* 22 (Apr. 1978): 86–101.

2. Hoover, "Public Law 273," 88; "Policy and Instructions Governing the Establishment of Sustained Yield Units," 21 July 1944, S-Plans, TM, Federal Units, 1936–1949, R-2, 56-A-144/62230, NARS-D.

3. Charles L. Tebbe to the files, 12 Aug. 1944, S-Plans, TM, Federal Units, 1936–1949, R2, 56-A-144/62230, NARS-D.

4. "Division of Timber Management 1945 Program of Work (Items to Be Stressed)," 21 Dec. 1944; and Ira J. Mason to the files, 29 Mar. 1945—both in DTMRF, RG95/1369.

5. Carter to the files, 11 Apr. 1945; and Granger to RF, Atlanta, 10 May 1945—both in DTMRF, RG95/1369.

6. Ira J. Mason to RF, Portland, 23 June 1945, DTMRF, RG95/1369.

7. Carter to RF, San Francisco, 4 July 1945, DTMRF, RG95/1369.

8. Watts to Col. Glenn L. Jackson, USAAF, 16 July 1945, DTMRF, RG95/1369.

9. Granger to RF, Portland, 18 July 1945, DTMRF, RG95/1369.

10. Granger to RF, Missoula, 27 July 1945; A. W. Greeley to the files, 31 July 1945; and Carter to RF, Portland, 1 Aug. 1945—all in DTMRF, RG95/1369.

11. "Hearing Record, Shelton Cooperative Sustained Unit, Hearing, September 19, 1946," TMO permanent files, PNWRO, FS. Except as otherwise indicated, the account of the establishment of the Shelton unit follows Hoover, "Public Law 273."

12. "Hearing Record."

13. Besides Hoover's excellent summary of the hundreds of pages of testimony and statements for the record, see the extensive file in the TMO permanent file, PNWRO, FS; see also Watts to the record, 10 Dec. 1946, and to Representative Walt Noran, 11 Dec. 1946, DTMRF, RG95/1368.

14. Watts to Magnuson, DTMRF, RG95/1368.

15. Granger to RFs, 23 Dec. 1946, DTMRF, RG95/1368. Hoover, "Public Law 273," 101, attributes this memorandum to Watts.

16. Watts to George H. Cecil, 16 Jan. 1947, DTMRF, RG95/1368.

17. Watts to Ellery Foster, International Woodworkers of America, 22 Jan. 1947, DTMRF, RG95/1367.

18. Watts to Region 1, 13 Mar. 1947, DTMRF, RG95/1367; Hoover, "Public Law 273," 101; Steen, *U.S. Forest Service,* 252. RG95/1367 and 1368 contain a lot of correspondence related to the futile attempts to establish cooperative sustained-yield units, most of which indicates utter lack of support in the communities to be blessed, as well as opposition from competing communities and industries, organized labor, small operators, and small holders. There are several recurrent themes. One was that opposition was widespread, from every conceivable type of person and organization. The Forest Service response was almost uniformly to blame opposition on "a lot of misinformation" (a frequent phrase) and similar aspersions. The Washington office repeatedly urged the regions to beat the bushes in the affected communities for support, to a point beyond futility.

19. Watts to RFs, 11 Apr. 1946, DTMRF, RG95/1368.

20. Granger to RF, Albuquerque, 15 May 1946, DTMRF, RG95/1369.

21. Steen, *U.S. Forest Service,* 252.

22. Granger to the record, 21 Jan. 1947, DTMRF, RG95/368; H. J. Andrews to RFs, all regions, 28 June 1949, S-Plans, TM, Federal Units, 1936–1949, R-2, 56-A-144/62230, NARS-D; and the following from TMO files, file 2410, Plans, PNWRO, FS: Sinclair A. Wilson, "A Preliminary Statement of Facts Bearing upon Forest Industries in the Grays Harbor Area, State of Washington," 16 Feb. 1948; Walter H. Lund to chief, 26 Apr. 1949; Watts to R-6, 24 May 1949; and USDA, FS, *Sustained Yield Timber Unit in Grays Harbor and Jefferson Counties* (typescript, transcript of hearings, 2 Aug. 1949).

23. "Establishment of the Proposed Big Valley Federal Sustained Yield Unit, Public Hearing, Alturas, California, October 19, 1949"; A. A. Hesel, "Amendment to Management Plan, Big Valley Working Circle," 15 Nov. 1949; and Harvey B. Mack, "A Study of the Big Valley Federal Unit, Modoc National Forest, 1950–1958"—all in TMO files, file 2410, Plans, California Regional Office, FS; Burton Clark and W. G. Charter, "A Study of the Big Valley Federal Sustained-Yield Unit, Modoc National Forest, 1959–1974," TSO files, file 2410, Plans, Coconino NF; "Information Concerning the Proposed Big Valley Federal Unit," 15 Aug. 1949—all in S-Plans, TM, Federal Units, 1936–1949, R-2, 56-A-144/62230, NARS-D.

24. David O. Scott to RF, 18 Sept. 1946; Duncan M. Lang to forest supervisor, Carson NF, 26 Sept. 1946; Scott, "Sustained Yield Case Study: Vallecitos Working Circle," 20 Mar. 1947; and L. W. Darby to chief, 20 Nov. 1953—all in file 2410, Plans, Vallecitos FSYU, SWRO, FS, Albuquerque.

25. See, e.g., Hartzer, *Half a Century,* 142, 144. In the 1960s, northern New Mexico saw an uprising of Spanish Americans who were resentful about several things, among them Forest Service policies.

26. Martinez to RF, 21 Nov. 1947, file 2410, Plans, Vallecitos FSYU, SWRO, FS.

27. DJK [Kirkpatrick] to Otto [Lindh], n.d. [1948], file 2410, Plans, Vallecitos FSYU, SWRO, FS; Ira J. Mason to R-3, 21 Jan. 1948; and Watts to R-3, 21 Jan. 1948—both in DTMRF, RG95/1367. In explaining why the Vallecitos unit was established, E. I. Kotok of the Forest Service told Senator Henry C. Dworshak that "this unit . . . is for the purpose of improving living conditions of a small and remote community of Spanish Americans" (10 Aug. 1948, RG95/1366).

28. Plan submitted with P. V. Woodhead to chief, 26 Jan. 1948, with revisions and amendments; R. E. McArdle to O. D. Connery, Vallecitos Lumber Co., 17 Sept. 1948; and McArdle to Jackson Lumber Co., 8 Oct. 1952—all in file 2410, Plans, Vallecitos FSYU, SWRO, FS; Lyle F. Watts to O. D. Connery, 31 Mar. 1948; and McArdle to Connery, 17 Sept. 1948—both in DTMRF, RG95/1366 and 1367.

29. L. A. Wall to the files, 23 Nov. 1952; and L. W. Darby to the chief, 20 Nov. 1953—both in file 2410, Plans, Vallecitos FSYU, SWRO, FS.

30. J. L. Jackson to C. Otto Lindh, 14 Jan. 1955; and Kirkpatrick to the files, 1 Apr. 1955—both in file 2410, Plans, Vallecitos FSYU, SWRO, FS; see also Vernon Bostick to Walter L. Graves, forest supervisor, 25 Apr. 1955, in the same file.

31. Buford H. Starky to Jackson Lumber Co., 2 Apr. 1955; W. L. Graves to Jackson Lumber Co., 30 Mar. and 18 Aug. 1955; Fred H. Kennedy to chief, 5 Apr. 1956; and Dahl J. Kirkpatrick, "Guide Lines to Be Used in Determining Compliance with Local Labor Requirements Imposed in the Harvesting and Manufacture of National Forest Timber from the Vallecitos Federal Sustained Yield Unit," 22 Apr. 1955, file 2410, Plans, Vallecitos FSYU, SWRO, FS.

32. Thomas M. Smith to Fred Kennedy, 9 Dec. 1955, file 2410, Plans, Vallecitos FSYU, SWRO, FS.

33. Dahl Kirkpatrick to Carson NF, 9 Feb. 1956; Kirkpatrick to J. L. Jackson, 9 Feb. 1956; Fred H. Kennedy to chief, 17 Feb. and 5 Apr. 1956; L. A. Wall to the files, 23 Feb. 1956; J. L. Jackson to Richard E. McArdle, received 21 Mar. 1956; Senator Dennis Chavez to Ezra Taft Benson, 27 Mar. 1956; Fred H. Kennedy to chief, 6 Apr. 1956; and McArdle to Jackson Lumber Co., 20 Apr. 1956—all in file 2410, Plans, Vallecitos FSYU, SWRO, FS.

34. Vallecitos SYU Hearing Record, petitions dated 20 Aug. 1956, file 2410, Plans, Vallecitos FSYU, SWRO, FS.

35. Catron & Catron to Sidney Williams, USDA, 18 Oct. 1956; and Richard E. McArdle, "Decision on Continuation of Vallecitos Federal Sustained Yield Unit, Carson NF, N. Mex.," 3 Jan. 1957—both in file 2410, Plans, Vallecitos FSYU, SWRO, FS.

36. Dahl J. Kirkpatrick to files, 1 May 1957; Edward C. Groesbeck to the files, 21 May 1957; Kirkpatrick to the files, 7 May 1957; and Richard E. McArdle to J. L. Jackson, 23 May 1957—all in file 2410, Plans, Vallecitos FSYU, SWRO, FS.

37. M. M. Nelson to R-3, 29 July 1966, 1440 Inspection, GFI TM, Mason Bruce, 11–29 Oct. 1965, 75-135/231179, NARS-LN; William D. Hurst to files, 9 Aug. 1967, 2410-Plans, box C-07-073-3-6, NARS-Fort Worth (NARS-FW).

38. B. H. Payne to RF, Albuquerque, 10 Aug. 1967; Don D. Seaman to RF, Albuquerque, 13 Mar. 1968; T. W. Koskella to forest supervisor, Carson NF, 27 May 1969; M. J. Hassell to RF, 15 May 1969; T. W. Koskella to William D. Hurst, 31 Oct. 1968; and Yale Weinstein, Duke City Lumber Co. to M. J. Hassell, 18 Apr. 1969, 2400-Plans, C-07-073-3-6, and 2400-Timber, FY 69, E-27-036-2-4, NARS-FW.

39. William D. Hurst to chief, 14 Feb. 1972; Edward P. Cliff to Yale Weinstein, Duke City Lumber Co. 4 Apr. 1972; and Weinstein to M. J. Hassell, 22 July 1977, file 2410, Plans, Vallecitos FSYU, SWRO, FS.

40. USDA, FS, *Coconino National Forest,* 12; A. A. McCutchen to chief, 9 Sept. 1947 (citing Carter of 12 May 1943); Duncan M. Lang to the files, 9 July 1946; James G. McNary to RF, 16 Sept. 1946 (application for sustained-yield unit); C. Otto Lindh to RF, 23 Oct. 1946; "A Case Study of the Flagstaff Federal Sustained Yield Unit," approved 9 Sept. 1947; and Roland Rotty to RF, 17 Feb. 1947—all in file 2410, Plans, TMO files, SWRO, FS.

41. C. Otto Lindh to Roland Rotty, 19 Feb. 1947; R. W. Hussey to TM, 8 Aug. 1947; "A Case Study of the Flagstaff Federal Sustained Yield Unit," approved 9 Sept. 1947; James G. McNary to Hon. Clinton P. Anderson, secretary of agriculture, 18 Sept. 1947; Anderson to McNary, 26 Sept. 1947; Lyle F. Watts to secretary of agriculture, 17 Oct. 1947; C. M. Granger to R-3, 20 Oct. 1947; and Roland Rotty to RF, 29 Oct. 1947—all in file 2410, Plans, TMO, SWRO, FS; L. S. Gross to I. J. Mason, 30 Oct. 1947, DTMRF, RG95/1367.

42. H. E. Ochsner to R-3, 1 Jan. 1948; Lyle F. Watts to Region 3, 5 Apr. 1948; and L. S. Gross to the record, 23 Apr. 1948—all in DTMRF, RG95/1366; USDA, FS, "In the Matter of Federal Sustained Yield Unit, Coconino National Forest, Memorandum of Southwest Lumber Mills, re Allocation of Timber Resources," received 10 Feb. 1948; James G. McNary, "Memorandum of Southwest Lumber Mills, Inc., re Allocation of Timber Resources," received 10 Feb. 1948; P. V. Woodhead to chief, 5 Apr. 1948; C. W. Granger to R-3, 5 Apr. 1948; Woodhead to G. R. Birkland and to James G. McNary, both 7 Apr. 1948; Roland Rotty to RF, 7 Apr. 1948; P. V. Woodhead to chief, 8 Apr. 1948; L. S. Gross to the record, 23 Apr. 1948; C. M. Granger to R-3, 29 Apr. 1948; Roland Rotty to RF, 7 May 1948; P. V. Woodhead to chief, 10 May 1948; C. M. Granger to R-3, 4 June 1948; G. R. Birkland to P. V. Woodhead, 29 June 1948; and Roland Rotty to RF, 7 May 1948 (for quotation)—all in file 2410, Plans, TMO files, SWRO, FS.

43. Granger to Region 3, 26 July 1948, DTMRF, RG95/95.

44. Duncan M. Lang, "Flagstaff Sustained Yield Unit under Public Law 273," 23 Oct. 1946; and P. V. Woodhead to the files, 25 Oct. 1946—both in file 2410, Plans, TMO files, SWRO, FS.

45. Watts to RF, Albuquerque, 25 Oct. 1948, DTMRF, RG95/1366.

46. Watts to RF, 1 Nov. 1948, DTMRF, RG95/1366. Watts was willing to give in to public sentiment in general on this issue, but he maintained that Flagstaff was an exception.

47. Watts to Region 6, 23 Nov. 1948, file 2410, Plans, TMO files, SWRO, FS.

48. "Resume of the Arguments for the Establishment of the Federal Sustained Yield Unit at Flagstaff," received 14 Dec. 1948; *Arizona Daily Sun* (Flagstaff), 1 Dec. 1948; *Arizona Labor Journal,* 6 Jan. 1949; *Arizona Farmer,* 8 Jan. 1949; R. T. Titus, Western Forest Industries Association, to P. V. Woodhead, 24 Jan. 1949; C. J. Warren, Southwest Lumber Mills, Inc. to C. Otto Lindh, 26 Jan. 1949; *Arizona Daily Sun,* 27 Jan. 1949; *Mail* (Winslow, Ariz.), 20 Jan. 1949—all in file 2410, Plans, TMO files, SWRO, FS.

49. "Transcript of Proceedings in re: Flagstaff Sustained Yield Unit, Public Hearing, Flagstaff, Arizona, February 2, 1949"; P. V. Woodhead by C. Otto Lindh to the chief, 22 Mar. 1949; Watts to Region 3, 11 Apr. 1949; Woodhead by George W. Kimball to R. T. Titus, Western Forest Industries Association, 17 Jan. 1949; Kenneth A. Keeney to RF, 24 May 1950; and Dahl J. Kirkpatrick by D. M. Lang to Coconino NF, 29 May 1950—all in file 2410, Plans, Flagstaff FSYU, TMO files, SWRO, FS.

50. Otto Lindh by L. W. Darby to the chief, 20 Nov. 1953, file 2410, Plans, Flagstaff FSYU, TMO files, SWRO, FS.

51. J. B. Edens, Southwest Lumber Mills (drafted by Dahl J. Kirkpatrick, FS), to forest supervisor, Coconino, 11 Jan. 1954; C. Otto Lindh by Kirkpatrick to J. B. Edens, 16 Feb. 1954; Ira J. Mason to Region 3, 11 Feb. 1954; *Arizona Daily Sun* (Flagstaff), 2 June 1954; J. M. Potter, Coconino Pulp and Paper, to K. A. Keeney, 26 Mar. 1956; Richard E. McArdle to Coconino Pulp and Paper, 30 Mar. 1956; Irving A. Jennings, Arizona Pulp and Paper, to McArdle, 22 June 1956; McArdle to Arizona Pulp and Paper, 22 June 1956; Clare Hendee for Edward P. Cliff to E. H. Weig, Ponderosa Paper Products, 14 Oct. 1964; and A. W. Greeley to Ponderosa Paper Products, 5 Aug. 1968—all in file 2410, Plans, Flagstaff FSYU, TMO files, SWRO, FS.

52. Dahl J. Kirkpatrick to the chief, 9 Nov. 1956, and enclosures; Fred H. Kennedy by Kirkpatrick to the chief, 14 Feb. 1957; and Kirkpatrick to the files, 24 July 1957—all in file 2410, Plans, Flagstaff FSYU, TMO files, SWRO, FS.

53. Southwest Lumber Mills, Inc., *Annual Report for the Year Ended April 30, 1957;* Fred H. Kennedy by Dahl J. Kirkpatrick to the chief, attn: Ira J. Mason, 22 Oct. 1957; R. W. Crawford to James M. Potter, 15 Jan. 1960; Kennedy by Kirkpatrick to M. E. Halfley, Arizona Development Board, 2 Feb. 1960; Kennedy by Kirkpatrick to the chief, 15 and 29 Feb. 1960; Richard E. McArdle by Edward P. Cliff to RF, Albuquerque, 15 Mar. 1960; Yale Weinstein, Duke City Lumber Co., to John T. Utley, 9 Nov. 1977—all in file 2410, Plans, Flagstaff FSYU, TMO, SWRO, FS.

54. J. Morgan Smith to Southwest Forest Industries, 13 Aug. 1962, file 2410, Plans, Flagstaff FSYU, TMO, SWRO, FS.

55. M. C. Galbraith to Kaibab Lumber Co., 15 July 1965; and statement, E. L. Quirk, vice-president, Southwest Forest Industries, 30 Sept. 1965—both in file 2410, Plans, Flagstaff FSYU, TMO, SWRO, FS.

56. R. M. Honsley to RF, 19 July 1965, file 2410, Plans, Flagstaff FSYU (Ponderosa Paper Products), TMO files, SWRO, FS.

57. C. T. Bunger to FS, 18 Apr. 1968; Phil Passalacqua Lumber Co. to FS, received 16 Apr. 1968; W. B. Finley to RF, 7 June 1968; A. T. Hildman,

Southwest Forest Industries, to Hurst, 10 Apr. 1969; and Hurst to Hildman, 22 Apr. 1969—all in file 2410, Plans, Flagstaff FSYU, TMO files, SWRO, FS.

58. T. W. Koskella for William D. Hurst to forest supervisor, 10 Mar. 1969; Rawleigh L. Tremain, USDA Office of General Counsel, to chief, FS, 23 Oct. 1963; Ralph F. Keobel, assistant general counsel, to chief, 3 May 1962; Washington office (no date or signature) to Region 3, 10 Mar. 1969; Don D. Seamon to RF, 21 Mar. 1969; F. Leroy Bond to chief, 28 Mar. 1969; Homer J. Hixon to M. M. Nelson, 29 Apr. 1969; B. H. Payne to Messrs. Greeley and Cliff, 9 May 1969; MMN [Nelson] to Cliff and Greeley, 12 May 1969; J. D. Porter, Western Pine Industries, Snowflake, Ariz., to Richard Worthington, 26 Mar. 1975; D. D. Westerbury for Worthington to J. D. Porter, 4 Apr. 1975; W. L. Evans to Porter, 20 Feb. 1975; R. E. Worthington to Joseph D. Cummings, Office of General Counsel, 18 July 1975; DTM to RF, R-1 through R-10; J. D. Porter, Western Pine Industries, to Jack Utley, Dec. 1975 (no date); and William L. Holmes to Porter, 6 Feb. 1976—all in file 2410, Plans, Flagstaff FSYU, TMO files, SWRO, FS.

59. F. Leroy Bond to the files, 17 Sept. 1969, citing memorandums from the Region to the chief, 22 Mar. 1949, and the chief to the Region, 11 Apr. 1949; William D. Hurst to the chief, 22 Sept. 1969; and Homer J. Hixon to RF, 5 Nov. 1969—all in file 2410, Plans, Flagstaff FSYU, TMO files, SWRO, FS.

60. "Periodic Reanalysis, Flagstaff Federal Sustained Yield Unit," 19 Nov. 1970; and F. Leroy Bond to the files, 17 Sept. 1969—both in file 2410, Plans, Flagstaff FSYU, TMO files, SWRO, FS.

61. R. E. Worthington to RFs, 5 Sept. 1975, TSO files, Coconino NF; James L. Matson, Kaibab Industries, to John T. Utley, 27 Oct. 1977; Gary F. Tucker, Southwest Forest Industries, to Jack T. Utley, 3 Jan. 1978; Yale Weinstein, Duke City Lumber, to Utley, 9 Nov. 1977; and J. D. Porter, Western Pine Industries, to Coconino NF, 27 Oct. 1977—all in file 2410, Plans, Flagstaff FSYU, TMO files, SWRO, FS.

62. Hassell by Gary E. Cargill to the chief, 27 Jan. 1978, file 2410, Plans, Flagstaff FSYU, TMO files, SWRO, FS; deposition of Michael A. Kerrick, 24 Oct. 1980, TSO files, Coconino NF.

63. Michael A. Barton for John R. McGuire to Hon. Bob Stump, House of Representatives, 15 Aug. 1978; Stump to McGuire, 28 July 1978—both in file 2410, Flagstaff FSYU, TMO files, SWRO, FS; "Periodic Reanalysis, Flagstaff Federal Sustained Yield Unit 1979"; "Summary of the Flagstaff Federal Sustained Yield Unit" (1979); and "Not for Public Distribution: Summary of Public Response Relating to the Periodic Reanalysis of the Flagstaff Fed. Sus. Yield Unit" (undated draft, probably 1980)—all in TSO files, Coconino NF.

64. Record of Public Hearing, 9 Jan. 1980, TSO files, Coconino NF.

65. Peterson to RF, Albuquerque, 20 May 1980; press releases 28 May 1980—both in file 2410, Plans, Flagstaff FSYU, TMO files, SWRO, FS.

66. "Policy Statement for Flagstaff Federal Sustained Yield Unit Policy," effective 1 Oct. 1981, submitted 12 Aug. 1980, file 2410, Plans, Flagstaff FSYU; and Order Granting Dismissal (Judge Valdemar A. Cordova), United States District Court for the District of Arizona, filed 20 Mar. 1981, and related correspondence, file 1570, Appeals, Southwest Forest Industries Dissolution of the Flagstaff Federal SYU Coconino NF, TMO files, SWRO, FS.

67. "Forester Acts on War Peril," *Oregonian* (Portland), 15 Aug. 1950; Steen, *U.S. Forest Service,* 252; Michael A. Barton for John R. McGuire to Hon. Bob Stump, House of Representatives, 15 Aug. 1978; and "Briefing for Chief

Peterson Regarding Flagstaff Federal Sustained Unit,'' n.d.—both in file 2410, Plans, Flagstaff FSYU, TMO files, SWRO, FS.

CHAPTER SIX. MULTIPLE USE, SUSTAINED YIELD, AND THE WINDS OF CHANGE

1. Cooperative Forest Management Act of 1950, 64 Stat. 473; Steen, *U.S. Forest Service,* 268; Edward C. Crafts, ''The Case for Federal Participation in Forest Regulation,'' speech at Yale Forestry Club and Yale Conservation Club, 5 Dec. 1951.

2. Hartzer, *Half a Century,* 194.

3. Steen, *U.S. Forest Service,* 268–69.

4. USDA, FS, *Information Digest,* no. 57, 2 July 1952.

5. Van [Fullaway] to Leo Bodine, 30 June 1952, NLMA records, box 28, USFS, FHS; Steen, *U.S. Forest Service,* 269–70.

6. W. C. Hammerle to H. C. Berckes, 19 July 1952, reporting on the 10 July meeting, NLMA records, box 28, USFS, FHS. There has been some contention over when and why McArdle let the issue drop. For various opinions see Steen, *U.S. Forest Service,* 270, and Hartzer, *Half a Century,* 194.

7. CHS [Saze?] to E. W. Tinker, 22 July 1953; Leo V. Bodine to J. M. Brown, 10 Aug. 1953; Bodine to A. Z. Nelson, 8 Sept. 1953; A. G. T. Moore to Bodine, 21 Aug. 1953; W. C. Hammerle to McArdle, 2 Sept. 1953; Nelson to Ernest L. Kolbe, 16 Oct. 1953; Kolbe to Bodine, 16 Oct. 1953; Nelson to Myron Krueger, 17 Nov. 1953; Bodine to McArdle, 24 Dec. 1953; Claire Engle to J. Earl Coke, 14 Jan. 1954; and Jim [James P. Rogers?] to Bodine, 12 July 1954—all in NLMA records, box 28, USFS, FHS; Steen, *U.S. Forest Service,* 270–71.

8. A. G. T. Moore to Leo V. Bodine, 2 Feb. 1953; Emanuel Fritz to Henry Bahr, 13 May 1953; George M. Fuller to NLMA members, 2 Dec. 1952; McArdle to Bodine, 14 Oct. 1955—all in NLMA records, box 28, USFS, FHS; Steen, *U.S. Forest Service,* 286–90; USDA, FS, *Timber Resources for America's Future,* Forest Resource Report no. 14 (Washington, D.C.: GPO, 1958). A recent study of projections of forest production and consumption is Michael Williams's ''Predicting from Inventories: A Timely Issue,'' *JFH* 28 (Apr. 1984): 92–98.

9. McArdle to RFs and directors, 2 July 1953, NLMA records, box 118, USFS—General, FHS; Steen, *U.S. Forest Service,* 290–95.

10. Granger to Congressman Charles E. Potter, 21 Jan. 1948, DTMRF, RG95/1367; *Proceedings of the Fourth American Forestry Congress, October 29, 30, 31, 1953* (Washington, D.C.: American Forestry Association, 1953), 24.

11. Steen, *U.S. Forest Service,* 300–301.

12. *Proceedings of the Fourth American Forestry Congress,* 162–64.

13. Ibid., 170–71.

14. RF, R-5, to Divisions of TM, FC, S&PF, Eng., CF & RES, and enclosures, 11 Mar. 1954, ''Chemical Control of Vegetation,'' 67-1099/224013, Forest Fire Lab, NARS-LN.

15. Steen, *U.S. Forest Service,* 301–3.

16. Bodine to A. Z. Nelson, 21 Feb. 1955, NLMA records, box 28, USFS, FHS.

17. NLMA, *Forest Policy Statement* (1956), NLMA records, box 28.

18. Steen, *U.S. Forest Service,* 303–4.

19. Hartzer, *Half a Century,* 188.

20. Steen, *U.S. Forest Service,* 304.

21. Ibid., 305.

22. Hartzer, *Half a Century,* 202–3; Edward P. Cliff, ''Outdoor Recreation and Multiple Use,'' speech before the Federation of Western Outdoor Clubs, 31 Aug. 1958, Edward P. Cliff Papers, vol. 1, FS History Section.

23. Richard E. McArdle (by Cliff and Mason), ''National Forest Timber Management and the Logging Industry,'' presented at the Forty-ninth Pacific Logging Congress, 12 Nov. 1958, Cliff Papers, vol. 1.

24. Charles A. Connaughton, *Multiple Use of Forest Land* (San Francisco: USDA, FS, Region 5, 1959); Edward P. Cliff, ''Multi-Use Management of the National Forests,'' 23 Mar. 1959, presented at the Twenty-first Intermountain Logging Conference, Cliff Papers, vol. 1.

25. Edward P. Cliff, ''The Program for the National Forests,'' *Southern Lumberman,* reprint from 15 Dec. 1959 issue; ''Roan—the Jewel of the Appalachians,'' presented at the Fourteenth Annual North Carolina Rhododendron Festival, 18 June 1960; ''Shorelines and Multiple Use,'' presented at the Thirty-eighth Annual Convention of the Izaak Walton League of America, 22–25 June 1960; and ''Multiple-Use Management on the National Forests of the United States,'' presented at the Fifth World Forestry Congress, Seattle, Washington, 29 Aug. to 10 Sept. 1960—all in Cliff Papers, vols. 1 and 2.

26. Multiple Use-Sustained Yield Act of 1960, 74 Stat. 215.

27. Edward P. Cliff, remarks at Fourth American Forest Congress, 30 Oct. 1953, Cliff Papers, vol. 1.

28. Edward P. Cliff, ''National Forest Timber Management Policies and Programs,'' delivered before the Intermountain Logging Conference, 26 Mar. 1956, Cliff Papers, vol. 1.

29. ''A Rough Draft, Short Biography'' of Edward P. Cliff, 3 Oct. 1957, Cliff Papers, vol. 1.

30. Edward P. Cliff, ''National Forest Transportation System,'' statement before Subcommittee on Roads, Senate Committee on Public Works, hearings at Albuquerque, N. Mex., 9 Dec. 1957, Cliff Papers, vol. 1.

31. Edward P. Cliff and C. A. Joy, speech before the annual meeting of the American Society of Range Management, 31 Jan. 1958; and Cliff, ''National Forest Transportation System,'' statement before Subcommittee on Roads, House Committee on Public Works, 4 Feb. 1958, Cliff Papers, vol. 1.

32. Cliff, ''Program for the National Forests.'' Industry spokesmen claimed that inadequate access to national-forest timber caused overcutting on private lands (see, for instance, George Craig, ''Public Timber Unused,'' *Pacific Logger,* June/July 1958).

33. McArdle (by Cliff and Mason), ''National Forest Timber Management.''

34. *Statistical History of the United States,* 540.

35. The following statistical discussion is derived from the tables in *Statistical History of the United States,* 534–41.

36. *Proceedings of the Fourth American Forest Congress,* 195–201.

37. For a summary of the technical distinctions between old-growth and young-growth timber and their use in construction see Clary, *These Relics of Barbarism,* 291–320.

38. ''Multiple Use Plan, Quilcene Ranger District, Olympic National Forest,'' 28 Jan. 1971, Appendix: ''Tentative Policies and Objectives Concerning Proposed Logging on the Big Quilcene Watershed,'' 6 Feb. 1955, PIO files, Olympic NF.

39. ''A Plan of Management for Timber Resources of Snoqualmie National Forest,'' 20 Dec. 1956; and Walter Lund by H. C. Hiatt (regional DTM) to supervisor, Siskiyou NF, 14 Jan. 1959, Historical Files, Siskiyou NF.

40. FS, *Manual,* 3, title 7, p. 9, chap. 1, ''Timber Use Policies,'' amended Dec. 1955.

41. In 1956 the Forest Service made the first comprehensive revision of the appraisal instructions since its last issue in 1922 and an incomplete attempt in the late 1930s. There was no substantive change, however (Cliff, ''National Forest Timber Management Policies and Programs'').

42. Edward P. Cliff, ''The Axe and the Saw,'' delivered before the Lake States Logging Congress, 15 Sept. 1956, Cliff Papers, vol. 1.

43. Edward P. Cliff, ''National Forest Transportation System,'' statement before Subcommittee on Roads, Senate Committee on Public Works, hearings at Albuquerque, 9 Dec. 1957; and ''National Forest Transportation System,'' statement before Subcommittee on Roads, House Committee on Public Works, 4 Feb. 1958; George Craig, ''Everyone's Future Is Tied to Forestry,'' speech to Society of American Foresters chapter, Yreka City, Calif., 10 May 1957, and ''Californians Need the Allowable Cut,'' speech to Sierra-Cascade Logging Conference, 14 Feb. 1958, courtesy Craig.

44. McArdle (by Cliff and Mason), ''National Forest Timber Management.''

45. Edward P. Cliff, ''The Care and Use of National Forests,'' *Yearbook of Agriculture 1958* (Washington, D.C.: GPO, 1958), 392–401.

46. By Cliff: ''Multiple Use Management of the National Forests''; ''Outdoor Recreation and Multiple Use''; ''Program for the National Forests''; ''The Role of Forest Recreation in Forest Land Management,'' address to meeting of the Washington Section, Society of American Foresters, 16 Feb. 1961; and by Donald J. Morriss: ''Timber Production on Recreation Areas,'' address to the Third Annual Meeting of the Washington Section, Society of American Foresters, 16 Feb. 1961—all in Cliff Papers, vols. 1 and 2.

47. ''Timber Management Plan, Rogue River National Forest, Oregon and California, Region 6, 1962,'' 12 June 1962, TMO files, file 2410, R6, TM, Plans, PNWRO, FS; Edward P. Cliff, ''Multiple Use Planning in National Forest Management,'' address to Third Annual Western Resources Conference, 8 Aug. 1961, Cliff Papers, vol. 2.

48. Western Pine Association, ''Minutes of Forest Conservation Committee,'' 21 Feb. 1952, NLMA records, box 28, USFS, FHS; Edward P. Cliff, ''National Forest Timber and Community Stability,'' panel discussion, Society of American Foresters, 31 Dec. 1953, and ''National Forest Timber Management.'' The Forest Service's devotion to an appraisal system devised in 1914 and not fundamentally revised since has brought ceaseless complaint from buyers. The system has also lost touch with technical and economic conditions in the industry (see George H. Craig, ''Forest Service Timber Appraisals,'' *JFH* 29 (Oct. 1983): 202–5.

49. George A. Craig, ''Californians Need the Allowable Cut,'' address to Sierra-Cascade Logging Conference, 14 Feb. 1958, courtesy Craig; McArdle by Clare Hendee to RFs, 18 Sept. 1958, NLMA records, box 129, USFS—Timber Sales Contracts, FHS.

50. By Edward P. Cliff: ''The Care and Use of the National Forests''; ''Recreation and Wildlife,'' dated 1953; ''Timber Accesss Road Problems on the National Forests,'' address to Pacific Logging Congress, 4 Nov. 1953; ''A 1956 Look at Timber Access Roads,'' address to Sierra-Cascade Logging Conference,

16 Feb. 1956; "National Forest Transportation System," statement before Subcommittee on Roads, House Committee on Public Works, 21 Feb. 1956; "National Forest Transportation System," statement before Subcommittee on Roads, Senate Committee on Public Works, hearings at Albuquerque, 9 Dec. 1957—all in Cliff Papers, vols. 1 and 2; see also B. H. Payne to the files, 13 May 1948, transmitting "An Access Road Program," DTMRF, RG95/1366; and NLMA, *Forest Policy Statement* (1956), NLMA records, box 28.

51. Remarks of H. L. Haller, Bureau of Entomology and Plant Quarantine, *Proceedings of the Fourth American Forestry Congress,* 184; and as an example of the latter problem, "Record of Review and Approval King's River North Working Circle Timber Management Plan," 15 Jan. 1962, TMO files, Sierra NF.

52. L. S. Gross to the record, 16 Feb. 1953, TMO files, file 2410, Plans, SWRO, FS.

53. FS, *Manual,* 3, title 7, chap. 1; Cliff, "Multiple-Use Management of the National Forests."

CHAPTER SEVEN. FROM MULTIPLE USE TO SUSTAINED PLANNING

1. Richard E. McArdle, "The Concept of Multiple Use of Forest and Associated Lands—Its Values and Limitations," address to the Fifth World Forestry Congress, quoted in Steen, *U.S. Forest Service,* 309.

2. As examples, see timber-management plans under constant revision during the early 1960s for the Sierra and Sequoia national forests in California, in the TMO files of those two forests.

3. "Timber Management Plan, Rogue River National Forest," 12 June 1962, TMO files, file 2410, R6, TM, Plans, PNWRO, FS.

4. Cliff by A. W. Greeley to RF, San Francisco, 26 Sept. 1962, TMO files, Plans, History, TM Statistics, Sequoia NF.

5. Homer J. Hixon by J. M. Buck to forest supervisors, 8 Nov. 1962, and Hixon by Fred Stallings to forest supervisors, 30 Jan. 1963, TMO files, Plans, History, TM Statistics, Sequoia NF.

6. Wilderness Act of 1964, 78 Stat. 890; David A. Clary, "Bureau of Outdoor Recreation" and "Public Land Law Review Commission," in Davis, ed., *Encyclopedia,* 1:54–55, 2:548–49; Steen, *U.S. Forest Service,* 311–14; see also Roderick Nash, *Wilderness and the American Mind* (New Haven, Conn.: Yale University Press, 1973), chap. 12.

7. Eastern Wilderness Act of 3 Jan. 1975, 88 Stat. 2096.

8. "Timber Management Plan, Willamette National Forest, Region 6—Oregon," 8 Mar. 1965, TMO files, file 2410, R6, TM, Plans, PNWRO, FS.

9. M. M. Nelson to RF, Albuquerque, 19 July 1966, 1440 Inspection, GFI-TM, Mason Bruce, October 11–29, 1965, R-3, 75-135/231179, NARS-LN.

10. Hurst to the record, 19 Dec. 1967, file 1360, Meetings, Staff, FY 1968, box 272528, NARS-FW.

11. Steen, *U.S. Forest Service,* 314–15; Michael Frome, *The Forest Service* (New York: Praeger, 1971), 86–87; Hartzer, *Half a Century,* 24, 252–55.

12. J. Phil Campbell, "Public Concern and Industry Interest in Forests," delivered before the American Pulpwood Association Annual Meeting, 18 Mar. 1970 (reprint, USDA, Office of the Secretary, 1970).

13. See, e.g., M. B. Doyle, Southwest Forest Industries, "Log, Roundwood, and Residual Utilization," address to the Society of American Foresters, 14 Oct. 1970, file 2400, TM, FY71, General, 75-1399/193108, NARS-LN.

14. United States Congress, Congressional Budget Office, *Forest Service Timber Sales: Their Effect on Wood Product Prices* (Washington, D.C.: GPO, 1980); current news reports, Feb.–Apr. 1983.

15. Congressional Budget Office, *Forest Service Timber Sales*, passim, quotation at xvii-xviii.

16. Clary, "Public Land Law Review Commission"; Steen, *U.S. Forest Service*, 320–21; U.S., Public Land Law Review Commission, *One Third of the Nation's Land* (Washington, D.C.: GPO, 1970). Two trenchant analyses of the commission's report, from different perspectives, are Michael McCloskey, "The Public Land Law Review Commission Report: An Analysis," Sierra Club *Bulletin* 55 (Oct. 1970): 18–30; and Edward P. Cliff, "Timber Resources," *Denver Law Journal* 54 (1977): 504–31.

17. Hartzer, *Half a Century*, 227–28. The following summary of RARE I and RARE II relies on the extensive collections of reports, clippings, and other materials on the subject in the History Section, FS, Washington, D.C.

18. For an example of that intention see Robert H. Torheim to forest supervisor, Mt. Baker-Snoqualmie, 3 Jan. 1975, TMO files, file 2450, Contracts and Permits, FY74-75, Roadless Areas, PNWRO, FS.

19. Steen, *U.S. Forest Service*, 311.

20. Rachel Carson, *Silent Spring* (Boston: Houghton Mifflin, 1962).

21. President's Science Advisory Committee, *Use of Pesticides* (Washington, D.C.: GPO, 1963); Steen, *U.S. Forest Service*, 318–19.

22. Steen, *U.S. Forest Service*, 319; Frank Graham, Jr., "Pest Control by Press Release: The Return of DDT," *Audubon* 76 (Sept. 1974): 65–71.

23. Department of Transportation Act of 1966, 80 Stat. 931; National Historic Preservation Act of 1966, 80 Stat. 915; National Environmental Policy Act of 1969, 83 Stat. 852.

24. The best summary of early litigation under NEPA is Harold P. Green, *The National Environmental Policy Act in the Courts (January 1, 1970–April 1, 1972)* (Washington, D.C.: Conservation Foundation, 1972).

25. Hartzer, *Half a Century*, 230.

26. Steen, *U.S. Forest Service*, 315–16; Hartzer, *Half a Century*, 21–22.

27. Ivan Doig, "The Murky Annals of Clearcutting," *Pacific Search* 10 (Dec./Jan. 1975–76): 12–14.

28. Carter to RF, Portland, 1 Aug. 1945, DTMRF, RG95/1369.

29. Mason to L. I. Barrett, 10 Apr. 1947, DTMRF, RG95/1367.

30. Ira J. Mason to Region 8, 14 May 1947; Mason to Region 1, 2 June 1947; C. M. Granger to Senator Glen H. Taylor, 10 June 1947; and Granger to Representative Abe McGregor Goff, 2 June 1947—all in DTMRF, RG95/1367; see also John R. Bruckart, "Taming a Wild Forest," *Yearbook of Agriculture 1949*, 326–34.

31. Mason to C. M. Granger, 1 Oct. 1947, DTMRF, RG95/1367.

32. Mason to C. M. Granger, 12 Nov. 1948, DTMRF, RG95/1366.

33. Bert Lexen, "Management of Lodgepole Pine," 21 Mar. 1949, S-Plans, TM, 1933–1950, R-2, 56-A-144/62230, NARS-D; John T. Auten and T. B. Plair, "Forests and Soils," *Yearbook of Agriculture 1949*, 114.

34. C. C. Buck and Russell LeBarron to W. A. Peterson, 7 Aug. 1953; and RF to Divisions, 11 Mar. 1954; "Chemical Control of Vegetation," 67-1099/224013; and S. E. Jarvi to RF, 22 Apr. 1953, S-Plans, TM, 59-543/91326, NARS-LN; Doig, "Murky Annals."

35. H. J. Gratkowski, ''Windthrow around Staggered Settings in Old-Growth Douglas-fir,'' *Forest Science* 2 (1956): 388–402; Gilbert H. Schubert, ''Early Survival and Growth of Sugar Pine and White Fir in Clear-cut Openings,'' California Forest and Range Experiment Station Publication no. 17, 21 Nov. 1956; R. R. Reynolds, *Eighteen Years of Selection Timber Management on the Crossett Experimental Forest,* USDA, FS Technical Bulletin no. 1206 (Washington, D.C.: USDA, FS, 1959).

36. Dr. William R. Halladay, Mountaineers Club of Seattle, statement, 6 Oct. 1961, *Use and Administration of National Forests,* hearings before the Subcommittee on Forests of the Committee of Agriculture, House of Representatives, 87th Cong., 1st sess. (Washington, D.C.: GPO, 1962), 57.

37. ''Timber Management Plan, Willamette National Forest,'' 8 Mar. 1965, TMO files, file 2410, R6, TM, Plans, PNWRO, FS.

38. USDA, FS, *Forest Service in a Changing Conservation Climate: I and E Approaches to Organizational and Public Needs 1966* (Washington, D.C.: USDA, FS, 1965).

39. The Forest Service appeared to be more inflexible and insensitive than it really was. For an example of quick action to correct technical mistakes see G. Lloyd Haynes to director, Rocky Mountains Forest and Range Experiment Station, 18 Oct. 1966, file 2400, Timber, folder #2, FY68, C-04-0770701, NARS-FW. On the other hand, it is worth observing how, since the 1950s, the Service has denied perhaps the most common public charge made against it—that it wants to increase the ''allowable cut'' on the national forests. Its response always has been a denial, to which is added that it wants only to raise the ''annual cut'' to bring it more into line with the ''allowable cut.'' By diverting the argument into a correction of the complainants' misinterpretation of Forest Service jargon, the Service apparently feels it has triumphed. Neither it nor most complainants seem to recognize that the Service has in the process admitted to the charge—what the complainants mean is the actual volume of harvest, and in wanting to raise the ''annual cut,'' that is what the Service says it wants to increase.

40. The principal recommendations of the five forestry-school deans are listed, together with the Forest Service's response as offered to the Council on Environmental Quality, in Forest Service Current Information Report no. 7, ''Forest Service Response to Recommendations of Forestry Deans,'' June 1972, copy in FS History Section, Washington, D.C.; see also Hartzer, *Half a Century,* 257.

41. Forest Service studies of the issue include *Management Practices on the Bitterroot National Forest* (Washington, D.C.: USDA, FS, 1970) and *Forest Management in Wyoming* (Washington, D.C.: USDA, FS, 1971).

42. David M. Smith, *The Practice of Silviculture,* quoted in Doig, ''Murky Annals.''

43. James B. Craig, ''Montana's Select Committee,'' *American Forests* 77 (Feb. 1971): 35–37, 48–49 (Bolle report); and Luke Popovich, ''The Bitterroot: Remembrance of Things Past,'' *JF* 73 (Dec. 1975): 791–93. For a contemporary foresters' defense of clearcutting see Kenneth P. Davis, statement before the Subcommittee on Interior and Related Agencies, Senate Committee on Appropriations, 17 Apr. 1970.

44. Edward C. Crafts, ''The Dilemma of the Forest Service,'' *American Forests* 76 (June 1970): 8, and ''Foresters on Trial,'' *JF* 71 (Jan. 1973): 14; Cliff's statement in FS's *Friday Newsletter* 34 (18 Sept. 1970): Hartzer, *Half a Century,* 23.

45. For Cliff's views see Hartzer, *Half a Century*, 189–91.

46. Doig, "Murky Annals"; Hartzer, *Half a Century*, 259.

47. USDA, FS, *National Forest Management in a Quality Environment: Timber Productivity* (Washington, D.C.: USDA, FS, 1971); Cliff to RFs and directors, 2 Jan. 1971, TMO files, file 2470, Silvicultural Practices, FY72, SWRO, FS.

48. *"Clearcutting" Practices on National Timberlands*, hearings before the Subcommittee on Public Lands of the Committee on Interior and Insular Affairs, Senate, 92d Cong., 1st sess. (1971); Hartzer, *Half a Century*, 24, 258.

49. See, for instance, Nancy Wood, *Clearcut: The Deforestation of America* (San Francisco: Sierra Club, 1971); Daniel R. Barney, *The Last Stand: Ralph Nader's Study Group Report on the National Forests* (New York: Grossman, 1974); Western Wood Products Association, "A Vital Aspect of Forest Ecology" (brochure, n.d., early 1970s); American Forest Institute, "Clearcutting: Instant Plunder or Resource Renewal" (brochure, n.d., early 1970s); *Arizona Republic* (Phoenix), 5 Mar. 1972, "Forest Stripping Embroils Conservationists"; Ladd S. Gordon, New Mexico Department of Game and Fish, to Michael Frome, *Field and Stream*, 26 Apr. 1972, file 2400, Timber, FY72, 75-131/193657, NARS-LN; W. Leslie Pengelly, "Clearcutting and Wildlife," *Montana Outdoors*, Nov./Dec. 1973, 26–30; and A. S. Harris, "Clearcutting, Reforestation, and Stand Development in Alaska's Tongass National Forest," *JF* 72 (June 1974): 33–37.

50. Hartzer, *Half a Century*, 272.

51. Forest and Rangeland Renewable Resources Planning Act of 17 Aug. 1974, 88 Stat. 476; USDA, FS, *A Summary of a Renewable Resource Assessment and a Recommended Renewable Resource Program* (Washington, D.C.: USDA, FS, 1976); Cliff, "Timber Resources," 507–31, offers a good summary of RPA requirements, based on official descriptions; see also USDA, FS, *Environmental Program for the Future* (Washington, D.C.: USDA, FS, 1974). For a history of the formulation of the RPA and the subsequent National Forest Management Act (discussed below) from the perspective of a congressional staff member who worked on them see Dennis C. LeMaster, *Decade of Change: The Remaking of Forest Service Statutory Authority during the 1970s* (Westport, Conn.: Greenwood, 1984).

52. Hartzer, *Half a Century*, 230–31.

53. Germane here are 30 Stat. 34–36 and 33 Stat. 628.

54. FS, *Manual*, 3, title 7, chap. 1, copy in files, Rogue River NF.

55. *West Virginia Division of the Izaak Walton League, Inc.*, v. *Butz et al.*, 522 F. 2d 945 (4th Cir. 1975), affirming 367 F. Supp. 422 (Northern District of West Virginia, 1973). The Bolle committee had been asked to consider whether the Service was violating the 1960 Multiple Use Act in its clearcutting. The committee thought so, but the grounds for a lawsuit seemed vague (Richard W. Behan, personal communication). Seizing on the 1897 legislation was the work of the Walton League's attorneys.

56. This summary of the background of the Monongahela case relies heavily on the report of Sidney Weitzman, "Lessons from the Monongahela Experience," an in-Service analysis based on interviews with Forest Service personnel, Dec. 1977, copy in the FS History Section.

57. National Forest Management Act of 1976, 90 Stat. 2849; see also Cliff, "Timber Resources."

NOTE ON THE SOURCES

Research on the history of the Forest Service, or on American forest history in general, should begin with a close study of two marvelous productions of the Forest History Society: Richard C. Davis's *North American Forest History: A Guide to Archives and Manuscripts in the United States and Canada* (Santa Barbara, Calif.: Clio Books, 1977) and Ronald J. Fahl's *North American Forest and Conservation History: A Bibliography* (Santa Barbara, Calif.: ABC-Clio, 1977). The guidance they provide to documentary sources and published literature is unequaled.

The standard administrative and political history of the Forest Service, a good starting point for a study of timber management within that agency, is Harold K. Steen's *The U.S. Forest Service: A History* (Seattle: University of Washington Press, 1976). Other useful general works are Herbert Kaufman's *The Forest Ranger: A Study in Administrative Behavior* (Baltimore, Md.: Johns Hopkins Press, 1960) and Glen O. Robinson's *The Forest Service: A Study in Public Land Management* (Baltimore, Md.: Johns Hopkins University Press, 1975).

Secondary literature that bears directly or indirectly on Forest Service timber management ranges over the history and administration of the public lands, timber resources and economics, and such special subjects as the history of fire control or significant controversies. Political histories of the United States and of its various public-interest controversies are also helpful in establishing the context. Special mention is due the Forest History Society's *Journal of Forest History,* which for nearly three decades has compiled a body of literature on important subjects that are mostly not addressed elsewhere. This study also benefited from the technical and philosophical literature of forestry and allied disciplines, including the periodicals *Journal of Forestry* and *American Forests.* Especially useful were the technical publications of the United States Department of Agriculture and of the Forest Service, including some historical studies, but mostly series on technical matters.

The majority of this history is founded upon documentary resources. First among them are the Records of the Forest Service, Record Group 59, in the National Archives, the various records generated by the Division of Timber Management and other parts of the Washington office being especially pertinent. Equally important are the accessioned and unaccessioned records of the Forest Service in the regional Federal Records Centers. Those holdings are matched by current and historical records collections that are still being maintained by the Forest Service at its many regional offices, experiment stations, and national forests. Of particular interest is the Buck Collection of documents on the history of timber management, which is maintained in the Pacific Southwest Regional Office at San Francisco, California. The Buck Collection represents a great deal of work in combing the archives, and its continued maintenance will be a valuable service to history and to the Forest Service. Also important is the collection of pertinent clippings that is maintained by the Public Information Office of the Pacific Northwest Regional Office, in Portland, Oregon.

Of other collections, particular notice must be paid to two. One is that of the Forest Service's History Section, in Washington, D.C., which has a significant holding of publications and other materials on all phases of the history of the Forest Service. The other—an exceedingly important source of information on the subject of this study—is the library and archives of the Forest History Society, in Durham, North Carolina (formerly in Santa Cruz, California), which contain an unsurpassed wealth of publications and important documentary and manuscript collections on every conceivable aspect of forest history.

INDEX